WEATHER FORECASTING

RED BOOK

FORECASTING TECHNIQUES FOR METEOROLOGY

Weather Forecasting Red Book
by Tim Vasquez

ISBN 0-9706840-6-1 / CS4LS

First edition / Third printing

Printed in the United States of America

Weather Graphics Technologies
P.O. Box 450211 Garland, TX 75045
(800) 840-6280 fax (206) 279-3282
Web site: www.weathergraphics.com
servicedesk@weathergraphics.com

Table of Contents

Introduction

A sound argument can be made that a sizable fraction of forecasters have lost their connection with the actual science of forecasting due to the explosion of observational and forecast data on the Internet. It seems that one only has to assemble the right charts to see the forecast outcome. However, expert chart *interpretation* is not meteorology; it is a skilled crapshoot that rewards anyone with good pattern recognition abilities on a day where the models excel. Expert *analysis and diagnosis*, and the application of *forecasting* fundamentals, are what makes a forecaster a true practitioner of scientific method. Their talents stand out during unusual and difficult weather situations – when the Palmer Divide threatens to produce orographic storms; when only a few hundred feet of warm air makes the difference between sleet and rain; when saturated ground at dusk stirs up concern over fog.

Unfortunately, I have noticed that it is easy to find interpretation information in books and on the Internet, but extremely difficult to find sage principles that augment the operational forecast method. Since 1998 I have been interested in writing a guide for operational use that encourages diagnosis of the atmosphere and blends it with *subjective* and *objective* forecast principles. That's what this book is for. It distills hundreds of techniques, methods, and even some useful rules of thumb. Some of these are in daily use, but others are sadly inaccessible, forgotten, or out of print (for example I came up bone-dry trying to locate the 1984 revision of Dvorak's groundbreaking hurricane intensity method!). This book is also meant to provide a highly useful and relevant set of conventions, standards, and fundamentals for *immediate use at a forecast desk.*

Furthermore, this book is meant to complement rather than replace my other titles. *Weather Forecasting Handbook* will remain my primary introductory guide to forecasting fundamentals, while *Weather Map Handbook* will serve as a benchmark guide to the Internet's weather product spectrum. *Weather Forecasting Red Book* ties them all together with emphasis on operational application.

For me, quality is always an ongoing process, so barring unforeseen circumstances this will not be the final edition. With this project at 300 pages I still find that I have used only a small fraction of the references and titles in my meteorological technical library. Unfortunately the print deadlines cannot wait any longer, so barring unforeseen circumstances a second edition should make an entrance sometime in 2007 or 2008. If you have any comments or complaints, have spotted a mistake, or want to suggest additional techniques, please visit weathergraphics.com and send an email. I'll be happy to hear from you.

Special thanks to Jim Tang, Simon Brewer, Brian Stertz, Jeff Snyder, Beau Dodson, David Wolfson, and Stan Rose, who all contributed some input to the convective forecasting section. Thanks also to Becky Klier who suggested a book title that I nearly used.

TIM VASQUEZ
September 2006

The art of analysis and forecasting is to assume as little as possible and, as far as possible, to base the forecasts on conclusions drawn from actual observations.

– SVERRE PETTERSSEN, 1941
Norwegian meteorologist

SECTION A

Observation

> I often say that when you can measure what you are speaking about, and express it in numbers, you know something about it; but when you cannot measure it, when you cannot express it in numbers, your knowledge is of a meagre and unsatisfactory kind.
>
> LORD KELVIN, 1889
> English physicist

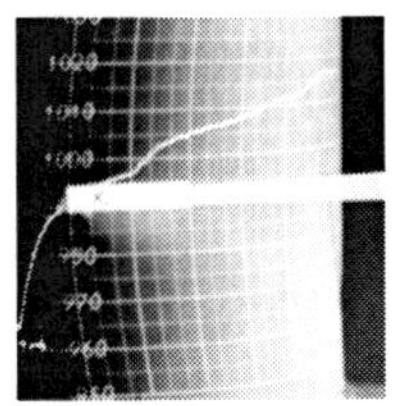

A1. OBSERVATION

Wind

Observation of wind in surface reports is, by convention, performed at the 10-meter (33 ft) level above ground in an open, exposed area.

1. Elements

1.1. **Wind direction** (*ddd*). In meteorology, wind direction always signifies *where the wind is coming from*, except when vectors are used (such as in trajectory forecasting). Wind is always referenced to *true north*. At the poles, north is oriented towards the prime meridian. The meteorological symbol for wind direction is *ddd*.

1.1.1. PERIOD. The wind direction is averaged over the previous *2-minute period* (FMH).

1.1.2. VARIABLE WIND. A variable wind direction is indicated when during this period it varies by 60 degrees or more when the average wind speed is greater than 6 knots. (FMH)

1.1.3. EXPRESSION AND UNITS. Wind direction is always expressed to the nearest ten degrees. A figure of "360" is used for north winds, and "000" is reserved for calm winds.

1.2. **Wind speed** (*ff*). Wind speed is defined as the rate at which air passes a given point.

1.2.1. PERIOD. Wind speed is determined by averaging the wind speed over the previous:

- *2-minute period*, in the United States (FMH)
- *10-minute period*, outside the United States

1.2.2. EXPRESSION. Wind direction is always *encoded* using two digits; possibly three depending on the report type if the wind value is 100 or greater.

1.2.3. UNITS. The unit of knots (nautical miles per hour) is always used in the United States, except in public forecasts where mph is required. Worldwide, knots or km/h are used. The unit of m/sec is used in some areas, especially throughout the former Communist bloc.

1.2.4. CALM WINDS. If there is no detectable motion of the air, the wind is considered to be calm. (FMH)

1.3. **Wind gust** ($f_m f_m$). A wind gust is the maximum instantaneous wind speed observed during a given time period. Its symbol is $f_m f_m$.

1.3.1. PERIOD. A wind gust is the maximum instantaneous wind speed that occurred the past *10 minutes*.

1.3.2. REPORTABILITY. A gusting condition is considered to exist if, during the observation period, there are rapid fluctuations in the wind speed with a variation of 10 kt or greater between peaks and lulls.

1.4. **Peak wind speed**. The maximum instantaneous wind speed measured is the peak wind.

1.4.1. PERIOD. Peak wind speed is determined for the period extending back to the last routine METAR observation is.

2. Characteristics

2.1. **Wind shift (U.S.)**. A wind shift is indicated when the wind direction changes by 45 degrees or more in less than 15 minutes with sustained winds of 10 kt or more throughout the wind shift.

2.2. **Wind squall**. A wind squall constitutes a sudden increase of wind speed by at least 16 kt, rising to 22 kt or more and lasting for at least one minute (FMH) (WMO Commission for Synoptic Meteorology, 1962). This may be associated with front or outflow passage.

2.3. **Gale**. A gale is defined as a surface wind with a mean speed of 34 kt or more (using the 10-minute average). (MET)

2.4. **Veering**. A clockwise change in wind direction with time or with height; e.g. south to west.

2.5. **Backing**. A counterclockwise change in wind direction with time or with height; e.g. south to east.

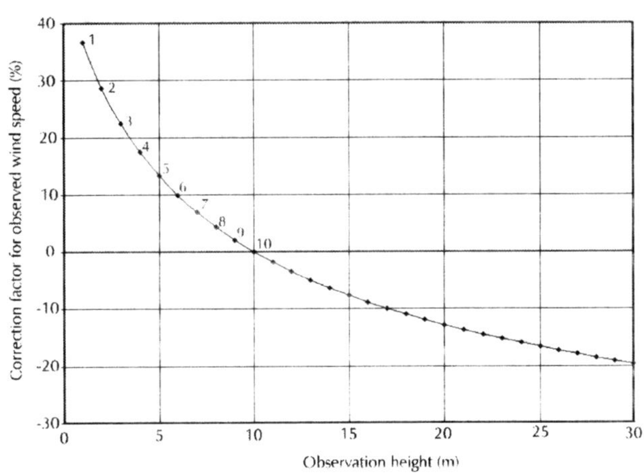

Figure A1-1. Correction of wind velocity at a particular observation height to the standard 10-meter height in international observational use. Designed for open exposures. Based on the UK Met Office formula in the text.

3. Miscellaneous

3.1. **Correction for height**. If wind speed cannot be determined at 10 meters, the following UK Met Office correction formula may be used to establish a correction. This is an approximation for open exposures.

$$V_h/V_{10} = 0.233 + 0.656 \log_{10}(h+4.75)$$

where V_h is the speed in knots at height h, V_{10} is the speed in knots at the 10-meter height, and h is the height in meters of the anemometer. For a convenient solution see the graph in Table A1-1. Note that this does not factor in obstacles such as trees and buildings, for which there is no reliable formula.

Figure A1-2. BEAUFORT WIND SCALE. The Beaufort Wind Scale was developed by Admiral Sir Francis Beaufort in 1805 to help sailors quantify wind measurements.

FORCE	SPEED		DESCRIPTION	EFFECTS
0	0-1 kt	0-1 mph	Calm	Calm; smoke rises vertically
1	1-3 kt	1-3 mph	Light air	Direction of wind shown by smoke drift, but not by wind vanes
2	4-6 kt	4-7 mph	Light breeze	Wind felt on face; leaves rustle; ordinary vanes moved by wind
3	7-10 kt	8-12 mph	Gentle breeze	Leaves and small twigs in constant motion; wind extends light flag
4	11-16 kt	13-18 mph	Moderate breeze	Raises dust and loose paper; small branches are moved
5	17-21 kt	19-24 mph	Fresh breeze	Small trees in leaf begin to sway
6	22-27 kt	25-31 mph	Strong breeze	Large branches in motion; whistling heard in telegraph wires; umbrellas used with difficulty
7	28-33 kt	32-38 mph	Near gale	Whole trees in motion; inconvenience felt when walking against the wind
8	34-40 kt	39-46 mph	Gale	Breaks twigs off trees; generally impedes progress
9	41-47 kt	47-54 mph	Severe gale	Slight structural damage occurs; chimney pots and slates removed
10	48-55 kt	55-63 mph	Storm	Seldom experienced inland; trees uprooted; considerable structural damage
11	56-63 kt	64-72 mph	Violent storm	Very rarely experienced; accompanied by widespread damage
12	64+ kt	73+ mph	Hurricane	--

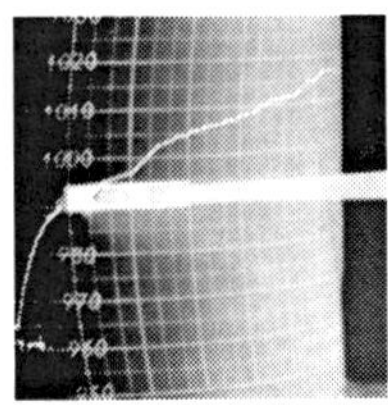

A2. OBSERVATION

Temperature and Moisture

1. Units

1.1. **Celsius** (Centigrade). Celsius is the dominant expression of temperature in meteorology work, with certain exceptions for Kelvin and Fahrenheit.

1.2. **Fahrenheit**. In meteorology, Fahrenheit is typically used only for surface observations in the United States.

1.3. **Kelvin**. Kelvin is used for expressions of potential temperature (theta) and equivalent potential temperature (theta-e). Values are expressed as Kelvin, not degrees Kelvin.

2. Expressions

2.1. **Ambient temperature** (T). Ambient temperature is measured in a sheltered spot about 2 meters above ground level.

2.2. **Dewpoint temperature** (T_d). Dewpoint temperature is the temperature at which water vapor, if cooled, would begin to condense. When it is below 0°C it is occasionally referred to as the *frost point.*

2.3. **Relative humidity** (q/q_s). The relative humidity is roughly equivalent to the ratio of the mass or vapor pressure of water in the air to the air's capacity for water vapor. A value of 100% indicates saturation, and a value of above 100% indicates supersaturation.

2.4. **Potential temperature** (theta, θ), K. Potential temperature is the temperature that results when a parcel is brought dry adiabatically to 1000 mb. Thus it can be used to negate the effect of elevation.

2.5. **Equivalent potential temperature** (theta-e, θ_e), K or °C. Equivalent potential temperature is the temperature that results when a parcel is brought wet adiabatically to 1000 mb. It can be used to compare parcel temperatures in a region against cap strength.

2.6. **Wet-bulb temperature** (T_w). This is the temperature that a parcel would have if cooled adiabatically to saturation by evaporation. Its value is normally between the temperature and dewpoint temperature. If the wet-bulb temperature is subfreezing, snow or hail will not melt even if the air temperature is above freezing.

2.7. **Wet-bulb potential temperature** (theta-w, θ_w), K or °C. The wet-bulb potential temperature is the wet-bulb temperature of a parcel moved adiabatically to 1000 mb. It has some value for hail forecasting.

2.8. **Mixing ratio** (w), g/kg. The mixing ratio indicates the amount of water vapor contained per unit of dry air. It is roughly proportional to dewpoint temperature.

2.9. **Saturation mixing ratio** (w_s), g/kg. The saturation mixing ratio indicates the capacity of water vapor per unit of dry air. It is roughly proportional to temperature.

2.10. **Vapor pressure** (e), mb. The vapor pressure indicates the pressure exerted by water molecules. It is rarely used in operational meteorology.

2.11. **Saturation vapor pressure** (e_s), mb. Saturation vapor pressure is the partial pressure that water vapor molecules would exert if the air were saturated. The ratio of vapor pressure to saturation vapor pressure equals the relative humidity.

2.12. **Specific humidity** (q), %. The ratio of the mass of the water vapor to the *total mass* of the parcel (including vapor).

3. Sultriness

3.1. **Temperature-Humidity Index (THI)**. In the United States this index was in widespread use until the late 1970s, when it was replaced by the Heat Index. The formula is:

$$T_Y = 0.4\,(T+T_w) + 15$$

where T_Y is the THI, and T and T_w are the temperature and wet-bulb temperature in deg F, respectively. A value of 72 or more indicates slightly uncomfortable conditions; 75 or more indicates acute discomfort; and 79 or more indicates general discomfort.

3.2. **Heat Index (HI)**. The Heat Index came into use in the United States during the early 1980s (Steadman 1979). A conversion table is most appropriate for manual conversions, however a mathematical approximation can be constructed from the following linear regression formula:

$$T_H = c_1 + c_2T + c_3R + c_4TR + c_5T^2 + c_6R^2 + c_7T^2R + c_8\,TR^2 + c_9\,T^2\,R^2$$

where T_H is the heat index (deg F), T is the ambient dry bulb temperature (deg F), R is the relative humidity (%), c_1 equals -42.379, c_2 equals 2.04901523, c_3 equals 10.1433127, c_4 equals -0.22475541, c_5 equals -6.83783×10^{-3}, c_6 equals -5.481717×10^{-2}, c_7 equals 1.22874×10^{-3}, c_8 equals 8.5282×10^{-4}, and c_9 equals -1.99×10^{-6}.

3.3. **Humidex**. The humidex was developed by Canadian forecasters (Masterton and Richardson 1979), and hinges on dewpoint rather than relative humidity. The formula is:

$$T_h = T + h$$

where h equals (0.5555)*(e - 10.0), e equals 6.11 * exp [5417.7530 * ((1 / 273.16) - (1 / T_d))], T_h is the humidex, and T_d is the dewpoint in *Kelvin*.

3.4. **Wet Bulb Globe Temperature (WBGT)**. The WBGT was developed by U.S. Air Force medical personnel to develop guidelines for hydration and rest; and it remains in use primarily in that field. Since it requires specialized equipment, its calculation is not discussed here.

3. Wind chill

Wind chill expressions attempt to combines the effects of low air temperature with the ability of air to remove the warm layer of air in contact with human skin. Most formulations are outgrowths of the 1945 Siple-Passel index (Siple and Passel 1945), which in turn is based on the time it takes for a vessel of near-freezing water to actually freeze under various wind and temperature conditions.

3.1. **Siple-Passel Wind Chill Index** (kcal/m², W/m²). The Siple-Passel index is the original expression of wind chill. Canada used the watts per square meter form from the 1970s until after 2000. Values of 1600 W/m² are associated with freezing of skin, with 2300 W/m² considered dangerous. The formula is:

$$Q = (10.45 + 10 \times \text{sqrt}(V) - V) \times (33 - T)$$

where Q is the rate of heat loss in kilocalories per square meter, Q_W is the rate of heat loss in watts per square meter, V is the wind speed in m/sec, and T is the ambient air temperature in deg C. To obtain W/m² multiply Q by 1.16.

3.2. **Wind Chill Equivalent Temperature** (°F). The Wind Chill Equivalent Temperature was in widespread use in the United States from 1973 until 31 October 2001. It is formulated as follows:

$$T_W = 0.0817\ (3.71V^{0.5} + 5.81 - 02.5V)\ \text{x}\ (T - 91.4) + 91.4$$

where T is the air temperature (deg F), and V is the wind speed (mph).

3.3. **New Wind Chill (JAG/TI) Index (°F)**. The new form of the Wind Chill Index (Figure A2-2) was adopted in Canada on 1 October 2001 and in the United States on 1 November 2001. It models wind speed at a height of 5 ft instead of the 33 ft height as reported, uses a human face model, incorporates modern heat transfer theory, lowers the calm wind threshold from 4 mph to 3 mph, uses a consistent standard for skin tissue resistance, and assumes no sun impact. It also corrects a problem with the artifically low numbers of the old index, which was used by the general public to suggest they had experienced much colder temperatures than they actually had. The new formula is:

$$T_W = 35.74 + 0.6215\text{T} - 35.75v^{0.16} + 0.4275Tv^{0.16} \quad \text{(imperial)}$$
$$T_W = 13.112 + 0.6215\text{T} - 11.37v^{0.16} + 0.3965Tv^{0.16} \quad \text{(metric)}$$

where T_W is the wind chill index (deg F or deg C), T is the air temperature (deg F or deg C), and v is the wind speed (mph or km/h).

3.4. **Steadman Wind Chill Index**. The Steadman Wind Chill Index was used in the United Kingdom until around 2001. It uses 22 criteria, including clothing variables, to determine heat loss. The formula is:

$$T_W = 1.41 - 1.162v + 0.980T + 0.0124v^2 + 0.0185vT$$

where T_W is the wind chill (°C), T is the temperature (°C), and v is the wind speed (m/sec).

Figure A2-1. National Weather Service manual observations are still taken at some locales, such as Valdez, Alaska (shown). *(Technician Wendy Zwickl pictured, courtesy Debra Russell)*

Wind (mph) \ Temperature (°F): Calm	40	35	30	25	20	15	10	5	0	-5	-10	-15	-20	-25	-30	-35	-40	-45
5	36	31	25	19	13	7	1	-5	-11	-16	-22	-28	-34	-40	-46	-52	-57	-63
10	34	27	21	15	9	3	-4	-10	-16	-22	-28	-35	-41	-47	-53	-59	-66	-72
15	32	25	19	13	6	0	-7	-13	-19	-26	-32	-39	-45	-51	-58	-64	-71	-77
20	30	24	17	11	4	-2	-9	-15	-22	-29	-35	-42	-48	-55	-61	-68	-74	-81
25	29	23	16	9	3	-4	-11	-17	-24	-31	-37	-44	-51	-58	-64	-71	-78	-84
30	28	22	15	8	1	-5	-12	-19	-26	-33	-39	-46	-53	-60	-67	-73	-80	-87
35	28	21	14	7	0	-7	-14	-21	-27	-34	-41	-48	-55	-62	-69	-76	-82	-89
40	27	20	13	6	-1	-8	-15	-22	-29	-36	-43	-50	-57	-64	-71	-78	-84	-91
45	26	19	12	5	-2	-9	-16	-23	-30	-37	-44	-51	-58	-65	-72	-79	-86	-93
50	26	19	12	4	-3	-10	-17	-24	-31	-38	-45	-52	-60	-67	-74	-81	-88	-95
55	25	18	11	4	-3	-11	-18	-25	-32	-39	-46	-54	-61	-68	-75	-82	-89	-97
60	25	17	10	3	-4	-11	-19	-26	-33	-40	-48	-55	-62	-69	-76	-84	-91	-98

Frostbite Times: 30 minutes; 10 minutes; 5 minutes

$$\text{Wind Chill (°F)} = 35.74 + 0.6215T - 35.75(V^{0.16}) + 0.4275T(V^{0.16})$$

Where, T= Air Temperature (°F) V= Wind Speed (mph)

Effective 11/01/01

Figure A2-2. New JAG/TI wind chill index in use in Canada and the United States since 2001.

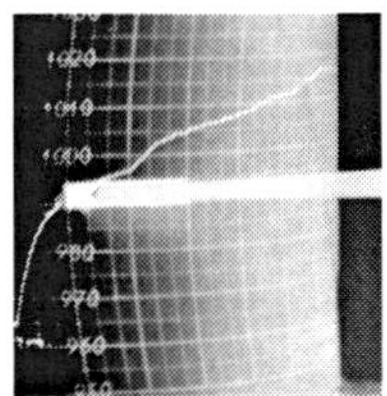

A3. OBSERVATION

Clouds

1. Cloud heights

1.1. **Expression**. In meteorology, the height of a cloud as determined by a reporting station always indicates the height of its *base*. This is expressed in feet or meters *above ground level*. Cloud tops, cloud heights obtained from a pilot report (PIREP), or thunderstorm tops use an expression of feet or meters *above sea level*.

1.2. **Etage**. The list below indicates the cloud *étage* ("level") spectrum, with typical heights in feet above ground level.

	Tropical	Mid-latitude	Polar
Low clouds	0-6500	0-6500	0-6500
Middle clouds	6500-26,000	6500-23,000	6500-13,000
High clouds	20,000-60,000	16,000-43,000	10,000-26,000

(WMO 1975)

1.3. **Ceiling**. The base of the lowest opaque cloud layer occupying *more than half of the sky* (broken or overcast) is considered the ceiling.

1.4. **Flight rules**. Ceiling and prevailing visibility establishes the aviation flight rules that are in effect. Instrument Meteorological Conditions (IMC) conditions require IFR (Instrument Flight Rules), while Visual Meteorological Conditions (VMC) permits Visual Flight Rules (VFR). In most countries:

- **IMC** is present when ceiling is below 1000 ft and/or visibility is less than 3 statute miles (5 km)
- **Marginal VMC conditions** are present when ceiling is below 3000 ft and/or visibility is less than 5 statute miles (8 km)

2. Cloud type

2.1. **Stratus** (St). Stratus appears as a low, uniform layer of clouds, resembling fog but not resting on the ground, and giving the sky a hazy appearance. It may appear in fragments beneath a precipitating cloud. It is rarely seen at heights above 2000 ft AGL.

- ❑ *WMO Type L6 if not associated with precipitation*
- ❑ *WMO Type L7 if associated with precipitation*

2.2. **Cumulus** (Cu). A well-defined cloud showing vertical development, and usually occurring with many other separate cumulus clouds. The upper surface is dome-shaped and does not appear fibrous. An exceptionally tall cumulus cloud is referred to as *towering cumulus* and may be transitioning into a cumulonimbus cloud. Cumulus clouds are often found at 2000 ft AGL in the morning hours, rising to 4000 ft AGL by afternoon; often much higher in desert regions.

- ❑ *WMO Type L1 for fair weather cumulus*
- ❑ *WMO Type L2 for towering cumulus*

2.3. **Stratocumulus** (Sc). A layer, or patches, of cloud composed of large globular elements or rolls. The cloud is usually gray with dark spots. Stratocumulus tends to form most frequently in water east of the subtropical cyclone and equatorward of the polar jet, and in cold advection behind a cold front (Weber 1981).

- ❑ *WMO Type L4 if originating from spreading of cumulus*
- ❑ *WMO Type L5 if not from spreading of cumulus*
- ❑ *WMO Type L8 if L5 and cumulus is at a different level*

2.4. **Cumulonimbus** (Cb). A large mass of cumuliform cloud, which resembles large mountains or towers when viewed from the side. The upper portions may appear as a veiled or featureless mass, called an anvil. Cumulonimbus produces lightning and thunder. Parts of the cloud where precipitation is falling generally appears hazy and featureless.

- ❑ *WMO Type L3 if no anvil is present*
- ❑ *WMO Type L9 if anvil is present*

2.5. **Nimbostratus** (Ns). A featureless low or middle cloud consisting of a dark gray sheet with precipitation falling from it.

The sun or moon is not visible through it. It is often obscured by precipitation and low-lying stratus. Nimbostratus is usually the result of altostratus which has thickened and lowered.

❑ *WMO Type M2*

2.6. **Altostratus** (As). A featureless middle cloud consisting of a sheet or veil, gray or bluish in color. The sun or moon may be completely obscured, and *light* rain may fall.

❑ *WMO Type M1 if transparent*

❑ *WMO Type M2 if opaque*

2.7. **Altocumulus** (Ac). A middle cloud consisting of well-defined layers or patches of cloud. Features within the altocumulus are often arranged in groups, lines, or waves.

❑ *WMO Type M3 if semitransparent*

❑ *WMO Type M4 if in patches*

❑ *WMO Type M5 if invading sky*

❑ *WMO Type M6 from spreading of cumulus*

❑ *WMO Type M7 opaque and not invading sky*

❑ *WMO Type M8 if castellanus species*

❑ *WMO Type M9 if from a chaotic sky*

2.8. **Cirrus** (Ci). A high cloud of delicate and fibrous appearance, and thin enough to not have shading except at sunrise and sunset. They may appear as hooks or tufts. The cloud is made of ice crystals. Thicker patches and sheaves are often debris originating from cumulonimbus.

❑ *WMO Type H1 if filaments or strands not invading sky*

❑ *WMO Type H2 if dense*

❑ *WMO Type H3 if from cumulonimbus*

❑ *WMO Type H4 if filaments invading sky*

2.9. **Cirrostratus** (Cs). This is a thin, almost featureless veil that readily produces halo phenomena. It often precedes rain or snow, and is rare compared to cirrus.

❑ *WMO Type H5 if close to horizon and invading sky*

❑ *WMO Type H6 if close to zenith and invading sky*

❑ *WMO Type H7 if completely covering sky*

❑ *WMO Type H8 if not covering or invading sky*

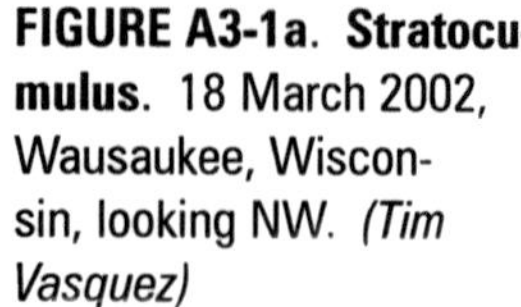

FIGURE A3-1a. Stratocumulus. 18 March 2002, Wausaukee, Wisconsin, looking NW. *(Tim Vasquez)*

FIGURE A3-1b. Cumulus in a warm weather regime.. The temperature was 93°F (34°C). 22 May 1999, Vernon, Texas, looking E. *(Tim Vasquez)*

FIGURE A3-1c. Cumulus in a strong cold advection pattern. Though cumulus is often associated with the summer season, the air temperature at this time was 9°F (-13°C). 21 March 2002, Dassel, Minnesota, looking W. *(Tim Vasquez)*

FIGURE A3-1d. Cumulonimbus developing in Arizona summer monsoon pattern. 18 July 1999, Ajo, Arizona, looking SW. *(Tim Vasquez)*

FIGURE A3-1e. Nimbostratus: a dull, featureless, opaque sheet that produces precipitation. A few shreds of stratocumulus lie underneath. 26 May 1999, Watson Lake, Yukon, looking W. *(Tim Vasquez)*

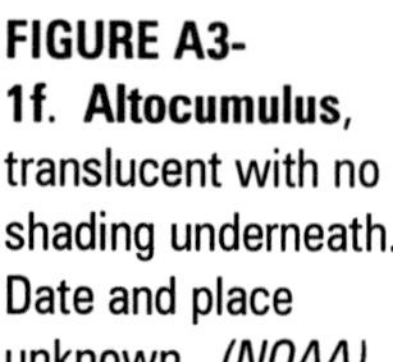

FIGURE A3-1f. Altocumulus, translucent with no shading underneath. Date and place unknown. *(NOAA)*

FIGURE A3-1g. Altocumulus in lenticular form observed in the lee side of the Rockies. Date and place unknown. *(NOAA)*

FIGURE A3-1h. Altostratus seen as a dull layer with the sun barely visible. Shreds of stratocumulus are in the foreground to the left of the rain gage. 26 October 2000, Norman, Oklahoma, looking SW. *(Tim Vasquez)*

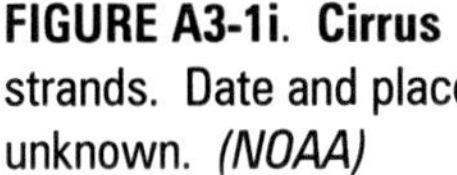

FIGURE A3-1i. Cirrus strands. Date and place unknown. *(NOAA)*

FIGURE A3-1j. Cirrus from the spreading of cumulonimbus anvils. 3 July 2001, Green River, Utah, looking NE. *(Tim Vasquez)*

FIGURE A3-1k. Cirrostratus showing a distinctive 22-degree halo. 5 June 2006, Norfolk, Nebraska, looking S. *(Laura Hedien)*

2.10. **Cirrocumulus** (Cc). Consists of cirrus containing very small globular or rippled elements, often giving rise to "mackerel sky". During daylight hours the elements do not have shading underneath. *(WMO Type H9)*

2.11. **Nacreous cloud**. Nacreous clouds are not accounted for by international meteorological networks. Observationally they are typically classified as cirrus.

2.12. **Noctilucent cloud**. Nacreous clouds are not accounted for by international meteorological networks. Observationally they are typically classified as cirrus or cirrostratus.

3. Cloud species

3.1. **Humilus**. Cumulus clouds of only slight vertical development. Commonly known as "fair weather cumulus". They are associated with weak instability. *Cu.*

3.2. **Congestus**. Cumulus clouds markedly sprouting and often of great vertical extent. Often known as "towering cumulus". They are associated with moderate or strong instability. *Cu.*

3.3. **Fractus ("scud")**. Appears as irregular, ragged shreds generally underneath a precipitating cloud. Coverage is usually scattered but they may consolidate into a broken or overcast layer and give the false impression of being the precipitating cloud. *St, Cu.*

3.4. **Lenticularis**. Highly striated lens or almond-shaped clouds. May be encoded as ACSL (altocumulus standing lenticular). Often associated with strong upper-level winds downstream of mountain ranges. May be an indicator of turbulence. *Cc, Ac, Sc.*

3.5. **Castellanus**. Clouds which are topped with cumuliform protuberances, and are often connected to a common base. Indicative of strong mid-level instability. *Ac, Cc, Sc.*

Figure A3-2a. Low clouds.

L1 Fair weather cumulus. Ragged Cu, other than bad weather, or Cu with little vertical development and seemingly flattened, or both.

L2 Moderate cumulus. Cu of considerable development, generally towering, with or without other Sc or Cu, bases at the same level.

L3 Cumulonimbus, no anvil. Cb, tops lacking clear outlines, but are clearly not fibrous, cirriform, or anvil-shaped; Cu, Sc, or St may be present.

L4 Stratocumulus spreading. Sc formed by spreading out of Cu; Cu may be present also.

L5 SC, not from spreading. Sc not formed by spreading out of Cu.

L6 Stratus of good weather. St in a more or less continuous layer and/or ragged shreds, but no fractus or bad weather.

L7 Stratus of bad weather. Usually underneath As or Ns.

L8 Cu & Sc, different levels. Cu and Sc (not formed by spreading of Cu); base of Cu at different level than base of Sc.

L9 Cumulonimbus with anvil. Cb having a clearly fibrous (cirriform) top, often anvil-shaped, with or without Cu, Sc, St, or scud.

Figure A3-2b. Middle clouds.

M1 Non-opaque altostratus. As, the greatest part of which is semitransparent through which the sun or moon may be faintly visible (ground glass).

M2 Altostratus/Nimbostr. As, the greatest part of which is sufficiently dense to hide the sun or moon, or Ns.

M3 Semitransparent Ac. Ac (mostly semitransparent) other than crenellated or cumuliform tufts; elements change slowly; bases at single level.

M4 Changing altocumulus. Patches of semitransparent Ac which are at one or more levels; cloud elements are continuously changing.

M5 Invading altocumulus. Semitransparent Ac in bands or Ac in one somewhat continuous layer, gradually increasing and spreading.

M6 Ac from spreading of Cu. Ac formed by the spreading out of Cu.

M7 Double layered altocumulus. Double layered Ac or an opaque layer of Ac, not increasing over the sky, or Ac coexisting with As or Ns or with both.

M8 Altocumulus castellanus. Ac with sprouts in the form of small towers or battlements or Ac having the appearance of cumuliform tufts.

M9 Chaotic altocumulus. Ac, generally at several layers in a chaotic sky; dense Ci is usually present.

Figure A3-2c. High clouds.

H1 Cirrus not invading sky. Ci in the form of filaments, strands, or hooks, not progressively invading the sky.

H2 Dense cirrus not from storms. Dense Ci, in patches or entangled sheaves, which usually do not increase.

H3 Dense cirrus from storms. Dense Ci, often in the form of an anvil, being the remains of the upper parts of Cb (cumulonimbus).

H4 Invading cirrus. Ci in the form of hooks or filaments, or both, progressively invading the sky.

H5 Cirrostratus low in elevation. Ci and Cs, or Cs alone; in either case, invading sky, but the continuous veil does not reach 45 deg above the horizon.

H6 Cirrostratus high in elevation. Ci and Cs, or Cs alone; in either case, invading sky, and continuous vail extends more than 45 deg above the horizon.

H7 Overcast cirrostratus. Veil of Cs covering the celestial dome.

H8 Cirrostratus not invading sky. Cs not progressively invading the sky and not completely covering the celestial dome. Ci and Cc may exist.

H9 Cirrocumulus. Cc alone, or Cc accompanied by Ci or Cs, but Cc is predominant.

Figure A3-3. Cloud amount expressions. Conversion of WMO oktas (eighths) to tenths. A trace of clouds is rounded up to 1 okta; a nearly-overcast layer is rounded down to 9 oktas.

Eights	Tenths	Descriptive
0	0/10	Clear
1 OKTA	1/10	Mostly clear/sunny
2 OKTAS	2/10 - 3/10	Mostly clear/sunny
3 OKTAS	4/10	Partly cloudy
4 OKTAS	5/10	Partly cloudy
5 OKTAS	6/10	Mostly cloudy
6 OKTAS	7/10 - 8/10	Mostly cloudy
7 OKTAS	9/10	Mostly cloudy
8 OKTAS	10/10	Cloudy
9 Sky obscured or cloud amount cannot be estimated		

Figure A3-4. NATO color codes. (Royal Air Force Flight Information Handbook)

COLOR	BASE OF LOWEST CLOUD LAYER OF 3/8 (SCT) OR MORE NOT LESS THAN	SURFACE VISIBILITY NOT LESS THAN
BLUE	2500 ft AGL	8 km (4.3 nm)
WHITE	1500 ft AGL	5 km (2.7 nm)
GREEN	700 ft AGL	3.7 km (2 nm)
YELLOW*	300 ft AGL	1.6 km (0.9 nm)
AMBER	200 ft AGL	0.8 km (0.4 nm)
RED	Below 200 ft AGL	< 0.8 km (0.4 nm)
BLACK	Airfield not usable due to non-weather reasons	
YELLOW 1*	500 ft AGL	2.5 km (1.4 nm)
YELLOW 2*	300 ft AGL	1.6 km (0.9 nm)

* At RAF airfields under the control of HQ 1 3 and at RN airfields.

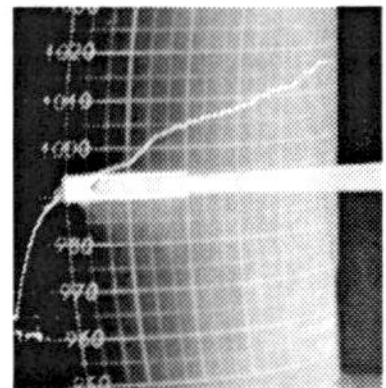

A4. OBSERVATION

Visibility

Visibility is a measure of the transparency of the atmosphere.

1. Units

Visibilities of less than 100 m (1/16th of a mile) are reported as zero.

1.1. **United States**. Visibility is measured in *statute miles* (one statute mile is 5280 ft or 1609 m).

1.2. **Outside of the United States**. Visibility is measured in *meters*.

2. Expressions

2.1. **Prevailing visibility** is the greatest distance that can be seen through at least half of the horizon circle, not necessarily through a continuous arc. When measured manually, a visibility chart with distances to known objects is consulted. With U.S. ASOS automated stations it represents a 10-minute sample of sensor outputs.

2.2. **Sector visibility** is the visibility in a specified direction that represents at least a 45-degree arc of the horizon circle.

2.3. **Runway visual range (RVR)** is the calculated maximum range at which a pilot on the runway can see the runway markings or lights. This is measured with *transmissometer* devices, consisting of a light beam near the runway transmitted to a detector 250 ft away. New technology is being deployed which measures forward scattering with a single device.

3. Flight rules

Ceiling and prevailing visibility establishes the aviation flight rules that are in effect. Instrument Meteorological Conditions (IMC) conditions require IFR (Instrument Flight Rules), while Visual Meteorological Conditions (VMC) permits Visual Flight Rules (VFR). In most countries:

- **IMC** occurs when ceiling is below 1000 ft and/or visibility is less than 3 statute miles (5 km)
- **Marginal VMC conditions** are present when ceiling is below 3000 ft and/or visibility is less than 5 statute miles (8 km)

A killer fog

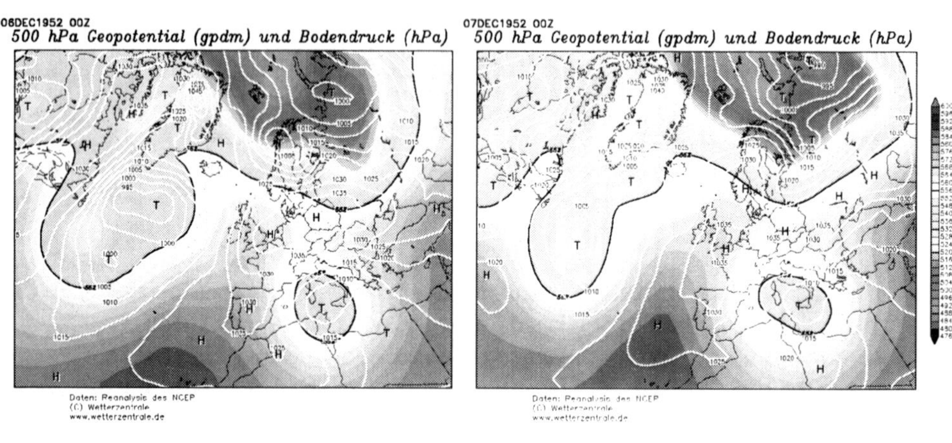

Figure A4-1. London's Killer Fog of 1952. Surface isobars (lines) and 500 mb upper-level heights (shading) on 6-7 December 1952. An extremely dense fog, laden with smoke, settled over southern England during this period. It claimed an estimated 12,000 lives. During the first few days of December 1952, the maps showed an omega block developing over the eastern Atlantic, locking up the patterns. A high pressure area stalled over the United Kingdom. Warm advection atop the stationary surface high led to a pronounced low-level inversion that trapped pollutants at the surface. Most of the pollution originated from the UK's dependence on coal, in this case, low-quality coal heavy with sulfur, which at the time enabled the UK to export its more valuable coal to ease an ongoing recession. The visibility was lowest on 7 December and deaths peaked on the 7th and 8th. The disaster led to major environmental reforms in 1956.

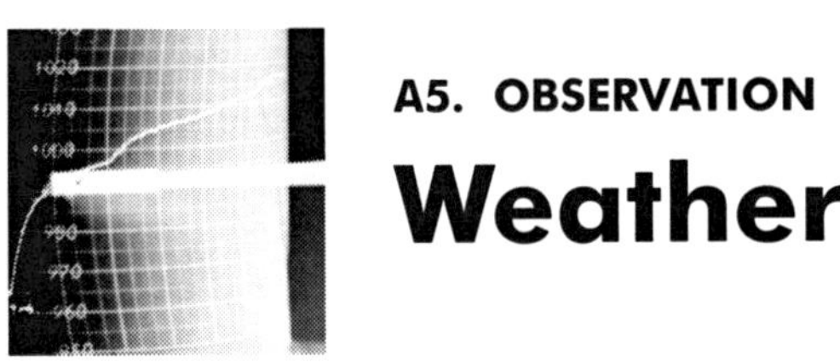

A5. OBSERVATION

Weather

In the mid-20th century, the World Meteorological Organization successfully standardized the definitions of different weather types. These definitions are recognized worldwide.

1. Weather

A hydrometeor is an atmospheric phenomenon composed of water.

1.1. **Rain** is composed of drops larger than 0.02 in (0.5 mm) in diameter. The drops do not appear to float or follow air currents. Light rain is comprised of clearly identifiable drops with no observable spray, and falls at the rate of up to 0.10 in/hr. Moderate rain falls at 0.11 to 0.30 in/hr, with drops not clearly identifiable and producing spray. Heavy rain falls at more than 0.30 in/hr, seemingly falling in sheets with heavy spray above pavement surfaces.

1.2. **Drizzle** is a uniform precipitation composed exclusively of fine drops of less than 0.02 in (0.5 mm) diameter. It appears to float when following air currents. Light drizzle does not reduce the visibility below 1/2 mile. Moderate drizzle reduces the visibility to greater than 1/4 mile but less than 1/2 mile. Heavy drizzle reduces the visibility to 1/4 mile or less.

1.3. **Snow** is composed of snow crystals, generally in the form of branched six-pointed stars. Light snow does not reduce the visibility below 1/2 mile. Moderate snow reduces the visibility to greater than 1/4 mile but less than 1/2 mile. Heavy snow reduces the visibility to 1/4 mile or less.

1.4. **Snow grains** are comprised of very small, white, opaque grains of ice. They form in cloud regimes that produce drizzle.

1.5. **Snow pellets** are made up of white, opaque grains of ice, usually round or conical, and with a diameter of 0.08 to 0.2 in (2 to 5 mm).

1.6. **Ice pellets**, or **sleet**, are precipitation of transparent or translucent pellets of ice. Light ice pellets do not completely cover an exposed surface regardless of duration. Moderate ice pellets create slow accumulation on the ground. Heavy ice pellets create rapid accumulation on the ground, and reduces visibility to less than 3 miles.

1.7. **Hail** is ice in the form of small balls or pieces of ice, falling individually or frozen together in irregular lumps.

1.8. **Thunderstorm** , in the United States, occurs when thunder is heard, or when lightning is observed at the station visually (with local noise levels drowning out thunder) or via automated means; the thunderstorm is noted as ending 15 minutes after the last occurrence of thunderstorm criteria.

1.9. **Shower** is precipitation that falls with a sudden beginning and end, changes intensity rapidly, and is accompanied by rapid changes in the sky.

1.10. **Freezing precipitation** occurs when the temperature is below freezing and liquid precipitation (rain or drizzle) is occurring, freezing and forming a glaze.

2. Obstructions to vision

2.1. **Mist** is a suspension of fine water droplets that reduces visibility to between 5/8 to less than 7 miles.

2.2. **Fog** is a suspension of fine water droplets that reduces visibility to less than 5/8 mile. The droplets do not fall to the ground; if they do, they are drizzle.

2.3. **Haze** is a suspension of extremely small, dry particles invisible to the naked eye. The particles may be made up of submicroscopic dust, soil, pollen, and fine pollutants. It casts a uniform veil over the landscape, subduing colors, and dark distant objects have a bluish tinge. Haze gives the sky a silvery tinge during midday, through which the sun appears dirty yellow. Sunsets have a reddish color.

2.4. **Smoke** is a suspension of small particles produced by combustion. If the smoke has travelled great distances it may transition to haze. Distant objects tend to appear dark gray or dark red. Smoke gives the sky a light gray or blue tinge, through which the sun appears orange. Sunsets tend to be very red.

2.5. **Volcanic ash** is a suspension of fine particles of rock powder that originate from a volcanic eruption. They may stay in the atmosphere for days or months.

2.6. **Dust** is comprised of fine particles of earth raised or suspended in the air by the wind. The sky has a pale, pearly, or tan color, the sun looks pale and colorless, and distant objects take on a tan or grayish tinge.

2.7. **Sand** is comprised of sand particles raised by the wind to a sufficient height to restrict horizontal visibility.

2.8. **Spray** is an ensemble of water droplets torn by the wind from the surface of an extensive body of water, generally from the crests of waves, and carried a short distance into the air.

3. Other weather conditions

3.1. **Dust whirl / sand whirl / dust devil / sand devil** is a whirling column of dust or sand particles, sometimes accompanied by small litter, raised from the ground.

3.2. **Tornado / funnel cloud / waterspout** is a violent, rotating column of air. A funnel cloud does not touch the earth's surface, and a waterspout forms over a body of water, touching the water's surface.

3.3. **Squall** (not to be confused with *squall line*) is a strong wind with a sudden onset during which the wind speed increases to at least 16 kt and is sustained at 22 kt or more for at least one minute.

3.4. **Sandstorm** is a weather condition comprised of particles of sand carried aloft by a strong wind. They are typically confined to the lowest ten feet above the ground, and rarely rise higher than 50 ft.

3.5. **Duststorm** is a weather condition comprised of fine particles of earth carried aloft by a strong wind. It may occur across an extensive area and rise to hundreds or thousands of feet in height.

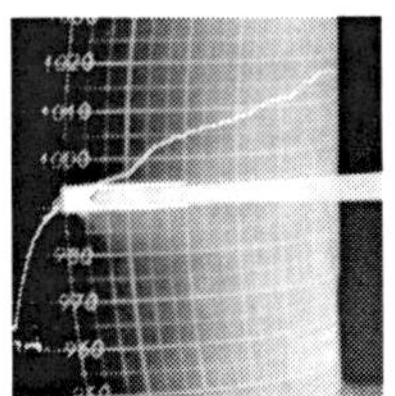

A6. OBSERVATION

Pressure

The abbreviation SI indicates the International System of Units. As this guide is intended for operational meteorology use rather than research work, millibars (hectopascals) will be used as the primary unit. This figure can be converted to kilopascals by dividing by 10. The pressure value in a standard atmosphere at sea level is given in brackets at the end of each paragraph.

1. Units

1.1. **Millibar (mb)**. Expression of atmospheric pressure in millibars is a de facto standard within operational meteorology, particularly in the United States. Although some experts argue that the millibar is a perfectly valid unit, its use is not generally accepted within research and academic environments, nor in scientific publication. *To convert millibars to kilopascals, divide by 10. Standard atmospheric pressure is 1013.25 mb.*

1.2. **Kilopascal (kPa)**. Kilopascals are regarded as the best SI measure for atmospheric pressure and are recommended for all formalized scientific work. It is the preferred unit of measure in American Meteorological Society publications. Kilopascals are also being accepted by the general public in countries such as Canada. *To convert kilopascals to millibars, multiply by 10. Standard atmospheric pressure is 101.325 kPa.*

1.3. **Hectopascal (hPa)**. Hectopascals, which are equivalent to millibars, were advocated as a way of expressing millibars in an SI framework. A considerable number of scientifically-minded individuals abhor the use of hectopascals, -hecto not being a preferred multiplier, with kilopascals being the recommended unit of measure. *To convert hectopascals to kilopascals, divide by 10. Standard atmospheric pressure is 1013.25 hPa.*

1.4. **Inch of mercury (in Hg)**. Atmospheric pressure was originally determined by measuring the height of a column of mercury. This is not an SI unit. However it is still in widespread use in the United States within aviation meteorology and by the general public. *To convert in Hg to mb, multiply by 33.8636. To convert in Hg to kPa, multiply by 3.38636. Standard atmospheric pressure is 29.92126 in Hg.*

2. Pressure measurements

Listed here are three types of pressure measurements commonly used. The corresponding phonetic abbreviation is given, used to identify the measurement in aviation and shipboard communications.

2.1. **Station pressure (QFE)**. The actual pressure being observed by the barometer, without reduction to sea level; e.g. what a mercury barometer would observe. This value is almost never transmitted in a surface weather observation. By convention it is always expressed in millibars, or inches or millimeters of mercury.

2.2. **Altimeter setting (ALSTG) (QNH)**. The result when the station pressure is reduced to sea level using an ICAO *standard atmosphere* (thus it is always the same regardless of the station location or the weather conditions). When this figure is dialed into an aircraft altimeter, it will read the aircraft's exact altitude above sea level. It is almost always expressed in inches of mercury to the nearest hundredth of an inch. The measurement is typically only used in air traffic control services and in public reports of "barometric pressure", but it may be used to create mesoscale surface analysis charts where reliable sea-level pressure values are scarce.

2.3. **Sea-level pressure (QFF)**. A theoretical expression of what the pressure should be if the barometer was moved to sea level. As a result, it has to factor in the estimated temperature of the air column below the ground. This can be a complicated process. In the United States, ASOS stations average the current temperature and the temperature 12 hours ago to eliminate diurnal effects, then extrapolate a column to sea level that has an assumed lapse rate of 6.5 K per km. Other stations may use a number of empirically-generated reduction methods. The result is almost always expressed in millibars, to the nearest tenth. This measurement is widely used to create surface analysis charts.

2.4. **Pressure altitude (PA)**. The pressure altitude figure is used only in aviation. It is the altitude in a *standard* atmosphere at which a particular station pressure will occur. It is also the altitude that an altimeter will show if it is dialed to 29.92 inches. A simple formula for figuring pressure altitude is:

$$A_p = H + (1000\,(29.92 - P))$$

where A_p equals pressure altitude in feet, H equals field elevation in feet, and P equals QNH pressure value in inches.

2.5. **Density altitude (DA)**. The density altitude is the pressure altitude corrected for temperature and humidity deviations from the standard atmosphere. A warmer temperature will result in less density and less lift. As this is a specialized and complex aviation figure, its computation will not be discussed here.

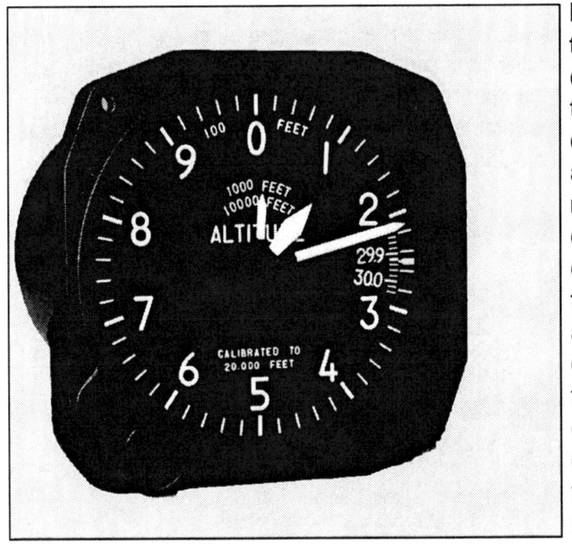

Figure A6-1. Altimeter. When the calibration knob is set to the observed altimeter setting (QNH), the altimeter will correctly read the actual elevation. When the calibration knob is set to 29.92 inches, the altimeter will read the observed pressure altitude. Each pressure change of 0.01 inch causes the altimeter value to change by about 10 ft.

ERVATIONS	LATITUDE	LONGITUDE	STATION ELEVATION(H_p)	TIME CONVERSION(LST to GMT) + Hrs - Hrs	MAG TO TRUE + Deg - Deg	DAY(LST)	MONTH	YEAR	STA & STATE OR COUNTRY
	15°11'N	120°33'E	+478 Feet(MSL)	-08	- 0	5	JUL	1982	CLARK AB LUZON R.P.

WEATHER AND OBSTRUCTIONS TO VISION LOCAL (5A)	LONG-LINE (5B)	SKY CONDITION (3)	TEMP (°C) (7)	DEW POINT (°C) (8)	ALSTG (Inches) (12)	REMARKS AND SUPPLEMENTAL CODED DATA (All times GMT. DESIRED ORDER OF ENTRY: Ceiling height, other remarks elaborating on preceding coded data, coded additive data group (if specified) radiosonde data, runway conditions, weather modification.) (13)	STA PRES-SURE (Inches) (17)	TOTAL SKY CVR (21)	OBS INIT (19)
		2SC035 3AC100 3AC150	24	22	2982	CIG100		8	X
		1ST001 5SC035 3AC100			2980	CIG035			X
FG	10BR	1FG/// 5ST001 5SC035	24	23	2980	CIG035 TWR VIS E 1/16 OCNL 3	29.290	8	X
FG	10BR	3FG/// 5ST001 5SC035			2980	CIG035 VIS NE1 TWR VIS E			X
						1/16 OCNL 3			
RA-FG	6IRA	2FG/// 5ST001 5SC035			2979	CIG035 VIS S1 TWR VIS 2			X
FG	10BR	2FG/// 5ST001 5SC035			2980	CIG035			X
FG	10BR	3ST001 3SC035 4AC100 1AC150	24	22	2980	CIG100		8	X
RA-	6IRA	1ST001 3SC035 3AC100 2AC150			2981	CIG100			X
	2IRERA	1ST001 2SC035 2AC100 2AC150			2981	CIG150			X
		1CI300							
	2IRERA	1ST001 3SC035 1AC100 2AC150	23	22	2982	CIG150		8	X
		2CI300							
		1ST001 2SC035 2AC100 2AC150	23	22	2981	CIG150	29.300	8	X
		2CI300							
RA-	6IRA	1ST001 1SC020 3SC035 1AC100			2982	CIG100			X
		2AC150 1CI300							
	2IRERA	1ST001 1SC020 1SC035 1AC100			2981	CIG150			X
		5AS150 1CI300							
RA-	6ORA	1ST001 1SC020 1SC035 1AC100	23	22	2982	CIG140		8	X
		5AS140 1CI300							
	2IRERA	2ST001 1SC020 1SC035 1AC100			2984	CIG140			RC
		5AS140 1CI300							
RA-	6ORA	2ST001 2ST002 1SC020 1SC035			2984	CIG035 CIGM002 OVR RWY			RC
		1AC100 3AS140							
RA-	6ORA	2ST001 2ST002 1SC020 1SC035	24	22	2985	CIG035 CIGM002 OVR RWY		8	RC
		1AC100 3AS140							

Figure A6-2. Pressure observations. The backbone of surface pressure observation in the United States is the station pressure (right column), a direct readout of pressure instruments. This is then converted to altimeter setting (left) using a set of tables, a circular calculator, or in the case of digital instruments, firmware. Sea-level pressure (not recorded here) is then calculated using the station pressure and the current temperature and the temperature 12 hours ago, using conversion procedures designed for the station.

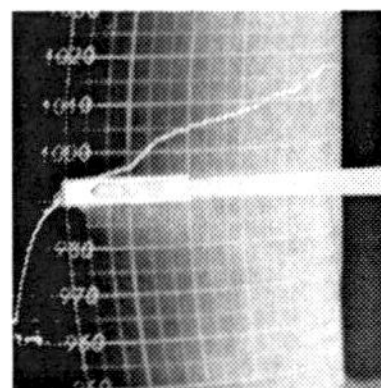

A7. OBSERVATION

Satellite imagery

Satellites are a rather new innovation to meteorology, first appearing during the early 1960s. Until the mid 1990s, availability was limited to large forecast offices and well-financed hobbyists due to the expense of display equipment. Weather satellite data is very much a cutting-edge technology with significant advances being made every several years.

1. Satellite types

1.1. **GOES (Geostationary Operational Environmental Satellite).** A geostationary satellite revolves around the earth at the exact same rate as the earth's rotation, so from the ground the satellite appears to be stationary (geosynchronous). Such a satellite is always above the equator at a distance of 22,300 miles (35,786 km). Weather satellites placed into this orbit provide the vast majority of the world's satellite-based weather products.

1.2. **POES (Polar Operational Environmental Satellite).** Polar orbiting satellites predate geostationary weather satellites. They rotate around the earth at a height of just a few hundred miles, and pass over the same location at the same time twice each day. They are valuable for specialized data collection, especially over polar regions.

1.3. MODIS (Moderate-resolution Imaging Spectroradiometer). This is a very new technology that arrived with the 1999 and 2002 launches of the Terra and Aqua spacecraft, respectively. They offer very high-quality imaging of the earth at 1 km resolution in 36 spectral bands, with some sensors detecting 250 m (820 ft) resolution. Products are available on the Internet in near-realtime. The drawback that the spacecraft are polar orbiters and imagery is less frequent than the GOES images. MODIS images can greatly augment case studies and reviews.

2. Product types

2.1. **Visible**. Visible satellite products provide daytime imagery of the Earth. Brightness corresponds to strong reflection (albedo) of sunlight. The higher and thicker the cloud, the brighter it will be; with the exception of morning and evening clouds and thunderstorm tops (Weber 1981). The visible channel is 0.52-0.75 μm, which is in the green-yellow-red part of the visible spectrum.

2.2. **Infrared**. Infrared imagery is used to provide 24-hour detection of cloud features. It is based on the law that every type of matter emits infrared radiation. Cold objects (such as high, or dense clouds) are given a bright color, and warm objects (such as low clouds and warm ground) are given a dark color.

3. Infrared imagery characteristics

The current generation of GOES satellites uses five primary infrared channels and one visible channel. These may be selectively combined using workstations and personal computers. In addition, *enhancement curves* may be used to change the grayscale that of a specific color palette.

3.1. **Infrared bands**. Following are infrared bands in common use. The bands selected here are those used by the GOES satellites, but the principles are useful for all image sources.

3.1.1. VISIBLE (0.65 μm) (GOES Channel 1). Visible channel. It depends on sunlight and reflection from areas with high albedo.

3.1.2. NEAR INFRARED (1.6 μm). Available on NOAA-AVHRR, and Terra and Aqua MODIS; it is nearest on the infrared spectrum to the visual color red. It is useful for distinguishing cloud cover and snow/ice cover.

3.1.3. NEAR INFRARED (3.9 μm) (GOES Channel 2). Useful for cloud phase and drop size. It is valuable for monitoring low clouds and fog. The characteristics of near IR are very similar to that of visible imagery.

3.1.4. INFRARED (6.7 μm) (GOES Channel 3). Water vapor channel. It detects the *absorption* of the earth's emitted spectrum by water vapor, rather than the actual presence of water vapor.

Thus it is most useful in warm climates and in the mid- and upper-levels of the atmosphere.

3.1.5. FAR INFRARED (10.7 μm) (GOES Channel 4). This is sometimes referred to as the "atmospheric window channel". It is the traditional infrared image used in routine forecasting and served as the infrared channel for pre-GOES-8 units. It is valuable for cloud top temperatures, thickness, and coverage.

3.1.6. FAR INFRARED (12.0 μm) (GOES Channel 5). Used for thin cirrus detection.

3.1.7. FAR INFRARED (13.3 μm) (GOES Channel 6). Used for thin cirrus detection.

3.2. **Channel combination**.

3.2.1. FOG IMAGE (10.7 μm - 3.9 μm). The difference in these two channels can be used to differentiate liquid water or ice. It is known as the "fog image" or "fog product", allowing detection of liquid water clouds at night that blend in with the ground on traditional infrared imagery.

3.2.2. CLOUD TOPS IMAGE (6.5 μm - 10.7 μm). This has been shown to identify cloud tops that are growing into dry tropospheric or stratospheric air.

3.2.3. GLACIATION IMAGE (13.3 μm - 10.7 μm). This can differentiate liquid cumuliform clouds from glaciated cumuliform clouds.

3.2.4. HOT SPOT IMAGE (10.7 μm + 12.0 μm - 3.9 μm). Used to detect fire hot spots.

3.3. **Enhancement curves**. Enhancement curves were used widely during the 1970s and 1980s to assist with the ambiguity of grayscale printouts. With the advent of personal computers, multicolor displays, and user-selectable palettes in the mid-1990s, their role has diminished. However some color schemes still are based on these predecessors.

3.3.1. ZA CURVE. The ZA curve simply provided greater contrast of a grayscale image by concentrating 40% of the available shades in the 5 to -60°C range.

3.3.2. MB CURVE. The MB curve was the most popular infrared enhancement curve. It introduced dark banding at -32.5°C and colder, allowing very cold tops, particularly those of thunderstorms, to stand out.

3.3.3. HF CURVE. The HF curve was developed for U.S. Pacific Coast forecasters, allowing examination of precipitation moving onshore. It is very similar to the MB curve but introduces more banding.

4. Imagery features

4.1. **Cloud patterns.** The observable cloud patterns on satellite can be differentiated into four primary types:

4.1.1. CLOUD SHIELD. A cloud shield is a broad cloud pattern which is not more than four times as long as it is wide; otherwise it is referred to as a *cloud band* (Weber 1981).

4.1.2. CLOUD BAND. A cloud band is a nearly continuous cloud formation which is more than four times longer as it is wide (Weber 1981). It is generally more than 60 nm wide.

4.1.3. CLOUD LINE. A cloud line is a narrow cloud band in which *individual cloud elements are connected* (Weber 1981). A line does not exceed 60 nm in width.

4.1.4. CLOUD STREET. A cloud street is a narrow cloud band in which *individual cloud elements are not connected* (Weber 1981). A street does not exceed 60 nm in width and most commonly is only a few miles wide. Cumulus streets, spaced several miles apart, often develop in benign regimes where there is little change in wind direction with height. The wind flow is parallel to the streets. The spacing between cloud streets is approximately 5 times the cloud top height.

4.2. **Closed cells**. Closed cell patterns (usually associated with stratocumulus) are associated with subsidence and anticyclonic low-level flow. They display a quilt-like appearance on visible imagery, and dull gray on infrared. They usually indicate surface winds of less than 25 kt. They may be associated with light precipitation and reduced visibilities. Behind a cold front, closed-cell areas are usually closer to the anticyclone than the cyclone.

4.3. **Open cells**. Open cell patterns (usually associated with cumulus) form when there is strong instability below a marine inversion, such as with cold advection above relatively warm waters. It appears as ringlets of cumulus with clear centers. They are associated with cyclonic low-level flow. Closed, doughnut-shaped cumulus patterns indicate wind speeds of less than 20 kt, while arc or elongated shapes indicate stronger winds [ATC/SAT]. In areas of open cell convection, the diameter of the cells is approximately 15 times the cloud top heights. Behind a cold front, closed-cell areas are usually closer to the cyclone than the anticyclone.

4.4. **Enhanced cumulus**. A small patch of open-cell cumulus which is more vertically developed (brighter) than surrounding clouds. It is often an immediate precursor to developing thunderstorms breaking an inversion. They are much more easily identified on high-resolution visible imagery.

4.5. **Rope cloud / arc cloud**. A thin line of cumulus; very common over oceanic regions. It may indicate an outflow boundary. These clouds are especially well-developed on the northern Australian coast, especially in September-November, where it is referred to as a *morning glory*.

4.6. **Transverse bands**. These are irregularly-spaced parallel bands of thin cirrus filaments and strands oriented *perpendicular* to the wind flow. They are caused by strong vertical wind shear near the jet stream.

4.7. **Baroclinic leaf**. The baroclinic leaf is a leaf-shaped cloud shield, spanning several hundred miles in width, which marks the initial developing stage of a baroclinic system. It normally has a shallow "S" shape on the sharp upstream edge of the cloud system.

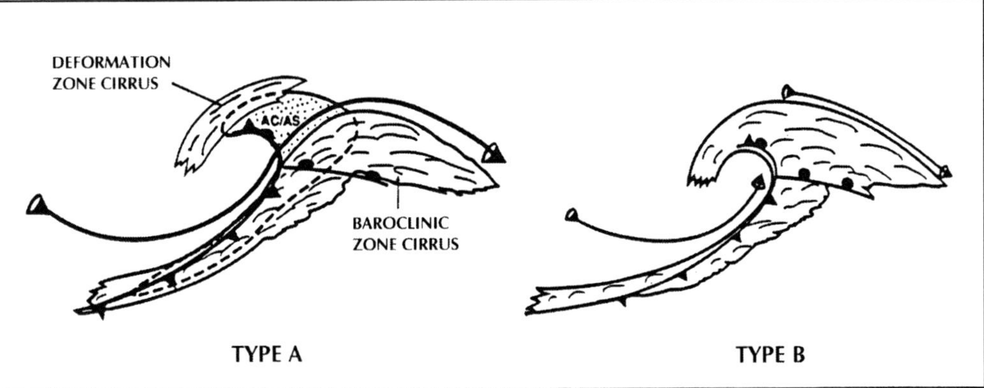

Figure A7-1. Conceptual model of baroclinic system clouds and jet in "Type A" and "Type B" baroclinic systems. [ATC/SAT]

The baroclinic leaf is usually found on the cold side of a frontal zone but is fed by warm air ascending the frontal surface.

4.8. **Baroclinic system**. Refer to Figure A7-1. The baroclinic cloud system is a very large cloud band associated with a frontal system that shows distinct cyclonic structure. It spans several hundred to thousands of miles.

4.8.1. TYPE A SYSTEM. The baroclinic cloud shield is discontinuous, with the upper-level low in a separate cloud mass. The jet is continuous, crossing the system at or just poleward of the triple point. There is a discontinuity between warmer clouds poleward of the jet and colder clouds equatorward; in visible imagery this looks like a transition between lumpy and smooth clouds, respectively.

4.8.2. TYPE B SYSTEM. The jet is discontinuous through the cloud system. The upper-level cloud band is continuous through its length. Latent heat release has caused the jet to fan out and wrap into the low.

5. Interpretation concepts

5.1. **Low clouds, snow, and ground**. Low clouds, snow, and ice often appear indistinguishable in visible and longwave infrared imagery. Animation can be of considerable assistance.

5.2. **Jet stream position**.

5.2.1. AXIS POSITION. The jet stream is located about 1 deg poleward of the sharp northern edge of a baroclinic zone cloud shield.

5.2.2. CLOUD ADVECTION. Where upper-level clouds are usually advanced downstream (or where the back edge is advanced furthest downstream), the jet intersects the cloud here.

5.2.3. OPEN/CLOSED CELL FORMATIONS. When no high clouds are present, the axis of the jet stream is normally located 1-3 deg poleward of the boundary between open cell cumulus and closed cell stratocumulus clouds.

5.2.4. WATER VAPOR DARK BANDS. Jet max positions are closely related to areas along the jet axis where there is a strong gradi-

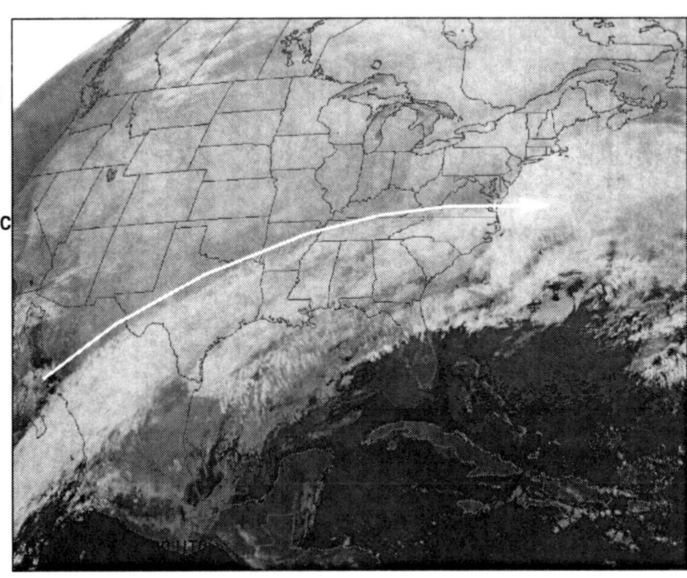

Figure A7-2. Subtropical jet stream (STJ) as seen on infrared imagery. The cloud shield shows the characteristic anticyclonic curvature and transverse banding.

Figure A7-3. Polar front jet (PFJ) as seen on back side of baroclinic system. Note that the jet divides open cell (OC) and closed cell (CC) regions of low-level cloud. The surface system is shown in black, and upper low in white (L), with a possible occlusion along the dashed-line.

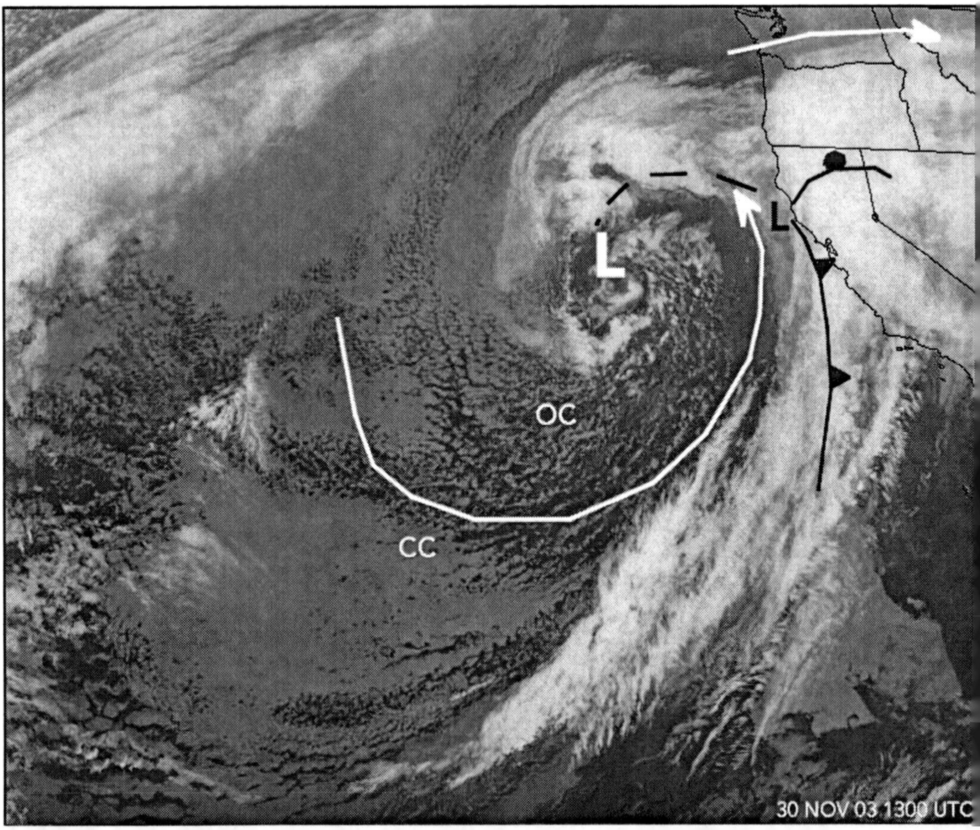

Figure 7-4a. Baroclinic leaf with mid-level shortwave trough (dashed line).

Figure 7-4b. Tropical cyclone. This image shows Hurricane Isabel off the Bahamas.

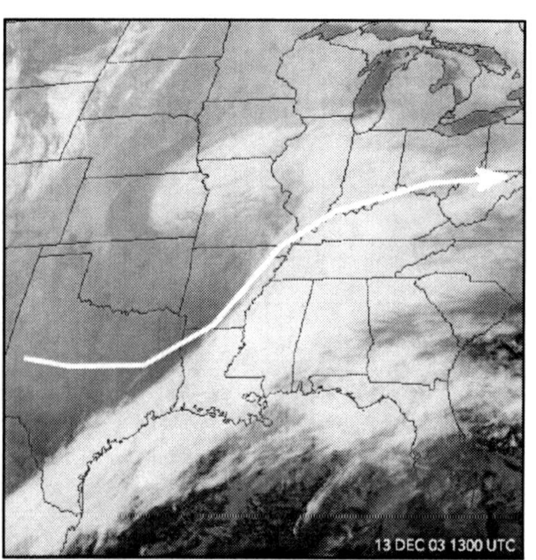

Figure 7-4c. Type A baroclinic low with jet crossing through system.

Figure 7-4d. Type B baroclinic low with jet wrapping into system.

Figure A7-5. Conceptual model for cloud systems in a cutoff low. (ATC/SAT)

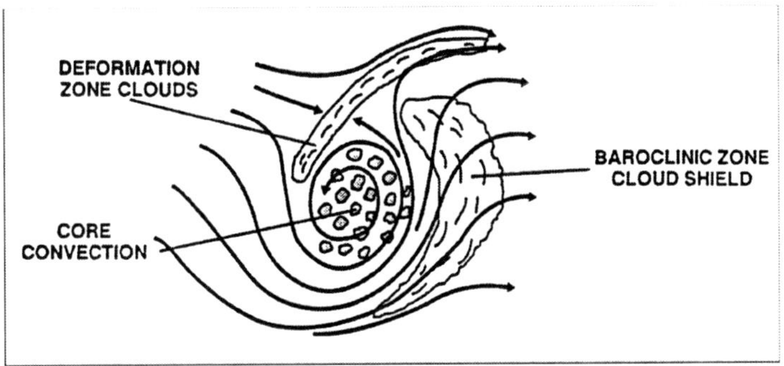

ent between dark and light signatures on water vapor imagery. The jet max is usually located in the dark band.

5.3. **Ridge axis**. High clouds are concentrated west of the upper-level ridge axis and may appear to "cut off" at the ridge axis. The lower the amplitude of the ridge, the more cirriform clouds will "spill" eastward past the ridge axis.

5.4. **Cutoff lows**. There are three basic cloud systems that may be found in a cutoff low (refer to Figure A7-2). The *baroclinic zone cloud shield* is associated with old baroclinicity ahead of the system. The *deformation zone clouds* are mid- and high-level clouds where the upper level flow elongates in the streamline deformation zone typically found poleward of the system. Finally the *core convection* is the most meteorologically significant area; it is associated with instability, and during the daytime may produce showers and thunderstorms.

5.5. **Estimation of sea level pressure**. The central pressure of a baroclinic cyclone may be estimated in lieu of any other data as follows. Refer to Figure A7-3:

- 990 - 1000 mb: The low has a well-defined "S" shaped pattern along the back edge of the frontal cloud band. The dry slot may begin to form when a low is at this intensity.
- 980 - 990 mb: A well-defined hook pattern is apparent in the cloud band. The dry slot is more apparent.
- 970 - 980 mb: A cloud band wraps once around the low center.

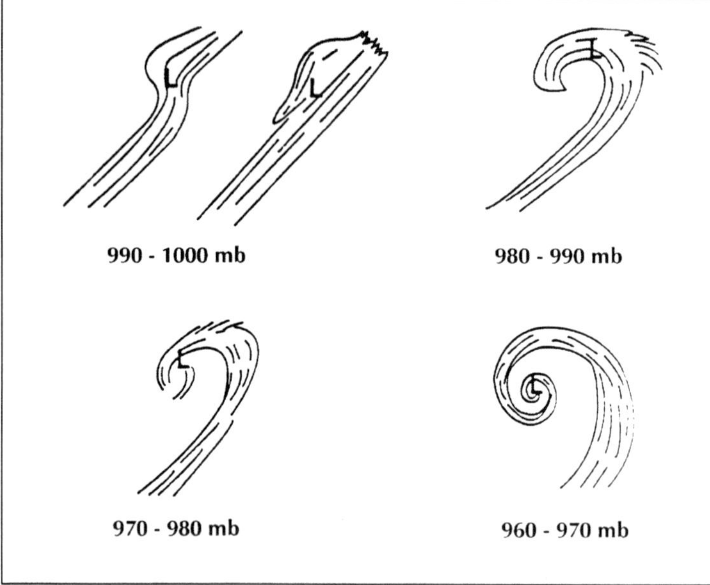

Figure A7-6. Estimation of extratropical cyclone central pressure based on satellite appearance. [ATC/SAT 1991]

- 960 - 970 mb: A cloud band wraps 1½ times around the low center.
- < 960 mb: Cannot be accurately estimated.

[ATC/SAT 1991]

5.6. **Water vapor imagery**. Water vapor absorption in the water vapor imagery band is most pronounced in the middle and upper troposphere (between 600 and 250 mb). The greater the radiation received, the darker the image. The brightness on water vapor imagery is a function of *both* water vapor content and temperature of the clouds and surface.

- *Warm low levels and moist upper levels will result in a bright pixel* (strong radiation from the low levels is absorbed in the upper levels and is detected as weak by the satellite)
- *Warm low levels and dry upper levels will result in a dark pixel* (strong radiation from the low levels is not absorbed in the upper levels and is detected as strong by the satellite)
- *Cold low levels and moist upper levels will result in a bright pixel* (weak radiation from the low levels is absorbed in the upper levels and is detected as weak by the satellite)
- *Cold low levels and dry upper levels will result in a bright pixel* (weak radiation from the low levels is not absorbed in the upper levels and is detected as weak by the satellite)

United States geostationary weather satellites

Name	Launched	Start Service	End Service	Notes
ATS-1	7 Dec 66	1966	1 Dec 78	
ATS-2	6 Apr 67	destroyed	destroyed	
ATS-3	5 Nov 67	1967	1 Dec 78	
ATS-4	10 Aug 67	destroyed	destroyed	
ATS-5	12 Aug 69	destroyed	destroyed	
SMS-1	17 May 74	1974	21 Jan 81	EAST; then moved
SMS-2	6 Feb 75	1975	5 Aug 82	WEST; then moved
GOES-A (1)	16 Oct 75	1975	7 Mar 85	EAST; moved 1976
GOES-B (2)	16 Jun 77	1975	1993	EAST; moved 1979
GOES-C (3)	16 Jun 78	1978	1993[3]	WEST; moved 1982
GOES-D (4)	9 Sep 80	1980	22 Nov 88	CEN; moved 1981
GOES-E (5)	22 May 81	1981	18 Jul 90[4]	EAST (CEN 1987-88)
GOES-F (6)	28 Apr 83	1983	12 Nov 94[5]	WEST (CEN 1984-87)
GOES-G	3 May 86[3]	destroyed	destroyed	Destroyed on launch
GOES-H (7)	28 Apr 87	1987	1996	EAST; moved 1989
GOES-I (8)	13 Apr 94	1994	3 Apr 03	EAST
GOES-J (9)	23 May 95	1995	27 Jul 98[2]	WEST
GOES-K (10)	25 Apr 97	27 Jul 98	online	WEST (repl. GOES-9)
GOES-L (11)	3 May 00	2000	[1]	CEN
GOES-M (12)	23 Jul 01	3 Apr 03	online	EAST (repl. GOES-8)
GOES-N (13?)	24 May 06	2006?	TBD	TBD
GOES-O (14?)	prob. 2006			
GOES-P (15?)	prob. 2007			
GOES-Q				
GOES-R	prob. 2012			

[1] Stored as full spare
[2] Stored as limited capability spare
[3] Still used as a communications relay satellite
[4] Major systems failure 30 Jul 1984
[5] Major systems failure 21 Jan 1989

GOES subpoint positioning by year

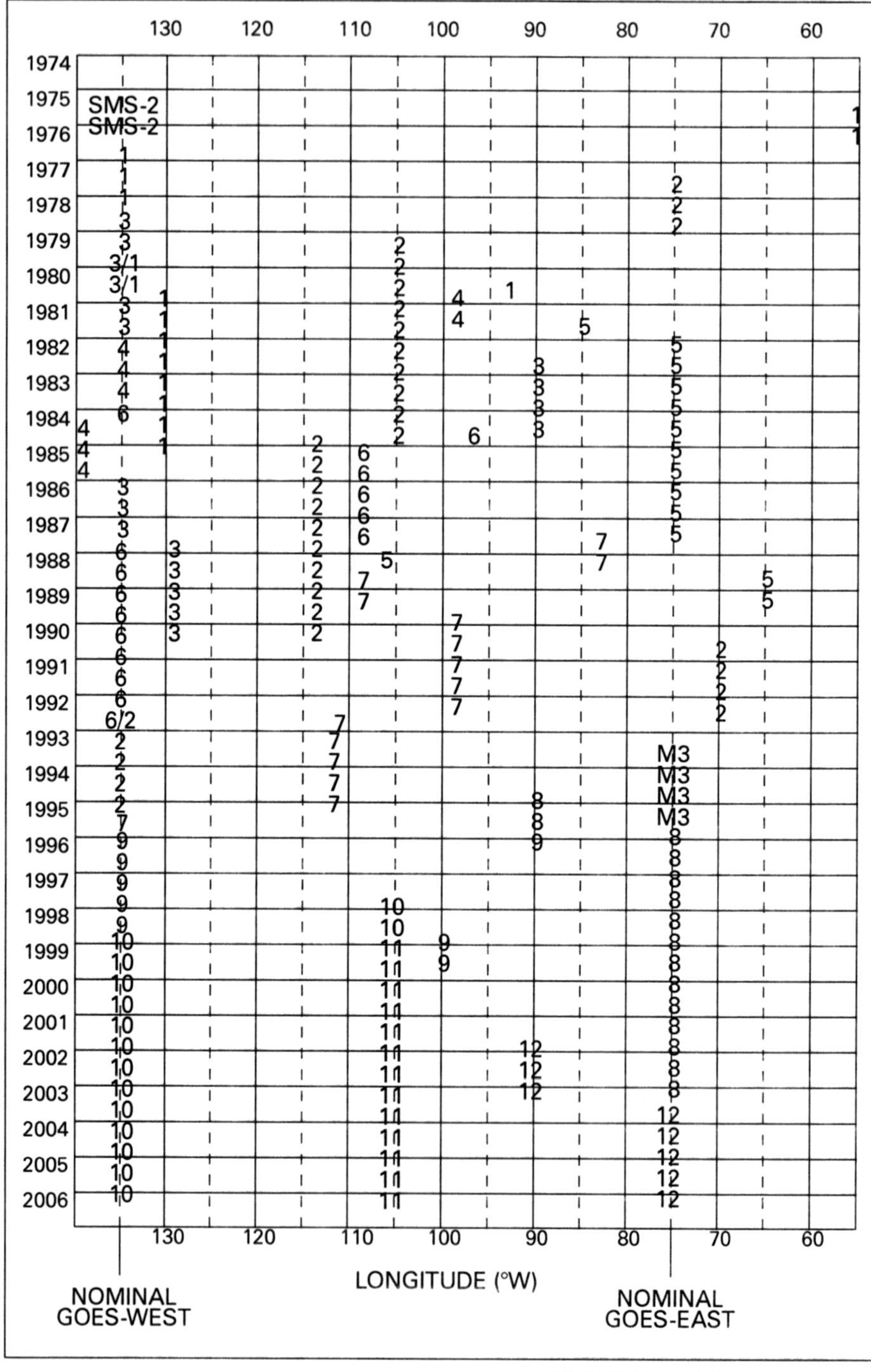

NUMBER = GOES NUMBER

SECTION B

Analysis

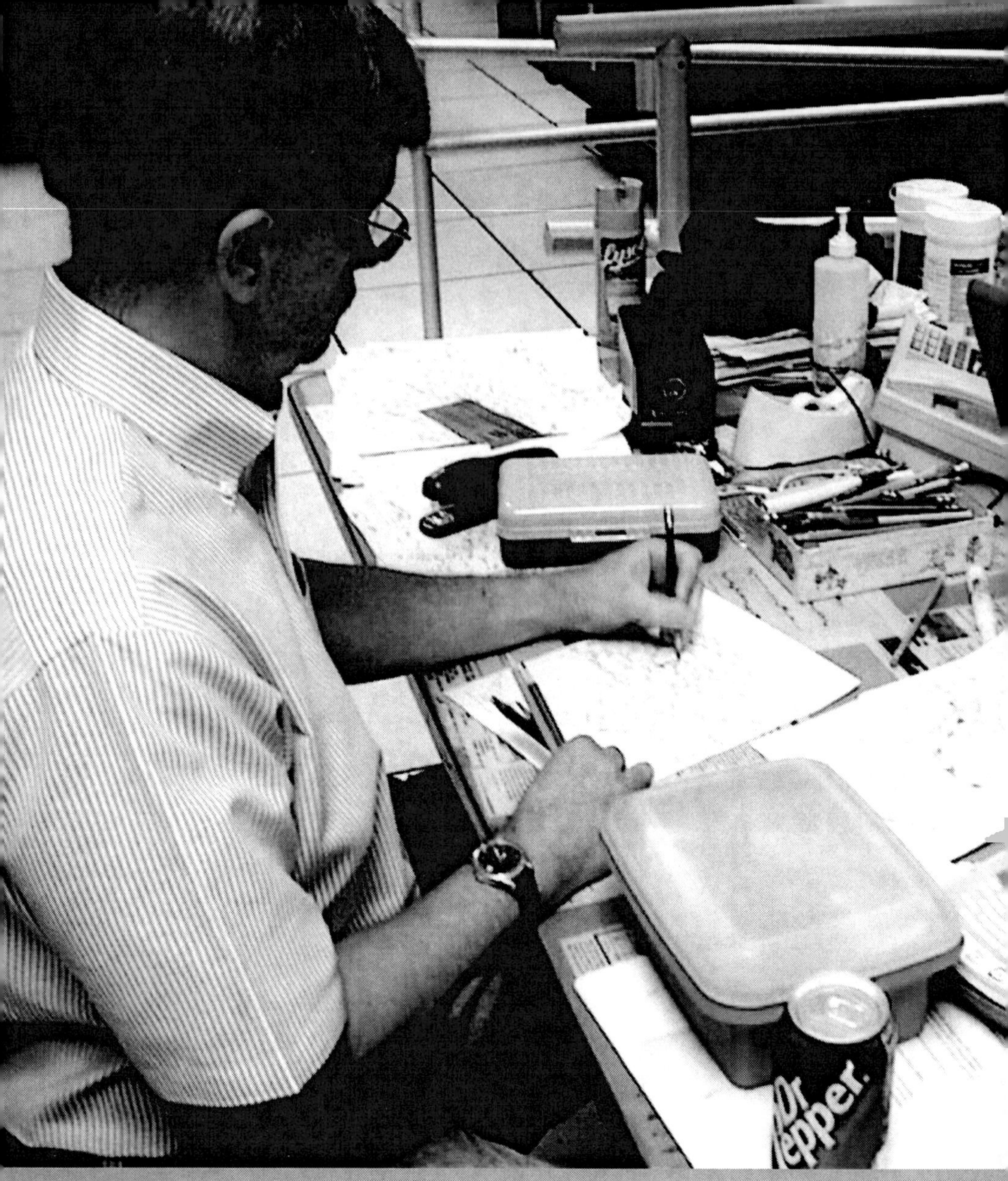

Facts do not cease to exist because they are ignored.

ALDOUS HUXLEY, 1927
British writer

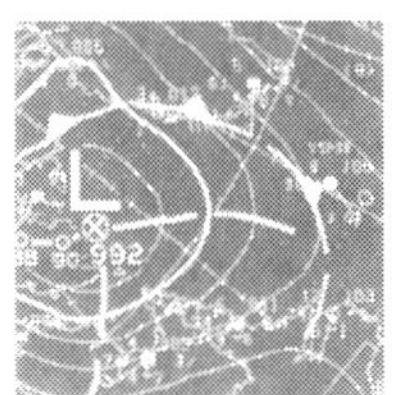

B1. ANALYSIS

Analysis principles

1. General

1.1. **Analysis**. Analysis the way a forecaster forms an understanding of what is happening in the atmosphere. Using one's meteorological knowledge to understand the analysis results in a *diagnosis*, which is a prerequisite to prognosis. (From Doswell 1986)

1.2. **Objective analysis**. Certain products may be quickly "solved" by using a computerized analysis, known as an *objective analysis*. A common form of objective analysis is computer-produced isobar charts.

1.3. **Anomalies**. A weather station may report conditions that differ substantially from those at surrounding stations. The data at that site either can be rejected as unrepresentative or treated like any other station's data (creating an unresolved "bump" in the resulting isolines). However, the best solution is for a human to employ an in-depth understanding of what caused those conditions to build a subjective model of those local events that comfortably incorporates all those "anomalous" data. (From Doswell 1986)

2. Hand analysis

2.1. **Sketching**. Work lightly in graphite pencil. Complete the entire analysis in pencil before beginning to harden in lines. Hardening may either be done boldly in pencil or in pen. [ATC/HDA]

2.2. **Isopleths**. Interpolate between station plots to estimate points where isopleths will pass through. Isopleths never cross, branch, or fork.

2.3. **Labelling**. Isobars should be labelled at the edges of the chart. Closed isobars should be labelled at the topmost part of each line. For English-language analysis, H should be used to identify points of high pressure, L for points of low pressure, W for warm points, K for cold points, X for other maxima, and N for other minima. The numeric values of these points should be written below the label.

2.4. **Boundaries**. Isopleths should never be extrapolated; that is, drawn outside of the data region. Lines should be terminated at the edge of the data and labelled.

2.5. **Closed isopleths**. The analyst should avoid drawing closed isopleths whenever possible, even at the occasional expense of creating long, narrow ribbons. There is good evidence that the atmosphere really does tend to create such features. (Doswell 1982)

B2. ANALYSIS

Standard levels

1. Geodetic levels

1.1. **Mean sea level** (MSL). Mean sea level is a reference height correlated to the mean height of the ocean. Heights above mean sea level are usually abbreviated AMSL or MSL. Most meteorological quantities are expressed relative to sea level.

1.2. **Ground level** (GL). In meteorology, heights above ground level are rarely used with the exception of *cloud heights* and *pre-NEXRAD radar heights*. Heights above ground level are abbreviated AGL. The sigma coordinate system (see Numerical Models) is used heavily in forecast models; this coordinate system uses the ground as its lowest level.

2. Pressure levels

Heights given are in geopotential feet and meters above mean sea level.

2.1. **200 mb**: *200 hPa / 20 kPa / 0.2 bar / 38662 ft / 11784 m*
Used for finding summer season jet streams in the temperate regions. This level may be part of the stratosphere, especially in the winter and poleward, and warm sinks may be found.

2.2. **250 mb**: *250 hPa / 25 kPa / 0.25 bar / 33999 ft / 10363 m*
Used for analyzing transition-season jet streams in the temperate regions.

2.3. **300 mb**: *300 hPa / 30 kPa / 0.3 bar / 30065 ft / 9164 m*
Used for finding winter season jet streams in the temperate regions.

2.4. **500 mb**: *500 hPa / 50 kPa / 0.5 bar / 18289 ft / 5574 m*
Located near the level of non-divergence (which averages near 550 mb). The 500 mb chart is typically used for finding short-wave disturbances in the large-scale flow.

2.5. **700 mb**: *700 hPa / 70 kPa / 0.7 bar / 9882 ft / 3012 m*
Used by severe weather forecasters to find cap strength.

2.6. **850 mb**: *850 hPa / 85 kPa / 0.85 bar / 4781 ft / 1457 m*
In non-mountainous terrain the 850 mb chart shows the upper part of the planetary boundary layer. It is used by severe weather forecasters in conjunction with soundings to help judge moisture depth and quality. In mountainous terrain the 850 mb level may be too close (or underneath) the ground.

2.7. **925 mb**: *925 hPa / 92.5 kPa / 0.925 bar / 2498 ft / 762 m*
Used by severe weather forecasters to assess moisture quality and analyze the low-level jet.

2.8. **1000 mb**: *1000 hPa / 100 kPa / 1 bar / 364 ft / 111 m*
Used as a substitute for sea-level pressure.

3. Flight levels

Flight levels are rarely used in non-aviation forecasting, however they may be encountered when working with pilot reports (PIREPs) and NCEP FDUS winds aloft output, the latter of which is being phased out. With the exception of the former Communist bloc, which uses whole meters, all countries express flight levels in *hundreds of feet, prefixed with FL* (e.g. 45,000 ft is FL450). A flight level does not use the ground or sea-level as its datum but rather the level where the pressure altitude equals zero. Thus a flight level will be closer to the ground in low pressure than in high pressure. Flight levels are not used in the lower troposphere. At these levels, altimeters are set to approximate actual altitude by using a barometric pressure value (altimeter setting) at a nearby station as a correction value.

4. Isentropic levels

The following levels are considered to be the best starting point for analyzing air masses as close to the ground as is practical. Refinement is required.

Season	Isentropic surface
Winter	290 - 295 K
Spring	295 - 300 K
Summer	310 - 315 K
Autumn	300 - 305 K

(Namias 1940)

When examining a constant pressure chart, the following chart may be used to select an appropriate isentropic surface based on the observed temperature.

	Isentropic level (K)									
	290	292	294	296	298	300	302	304	306	308
H850	4	6	8	10	11	13	15	17	18	20
H700	-11	-10	-8	-6	-4	-2	0	1	3	5
H500	-35	-34	-32	-30	-29	-27	-25	-24	-22	-21

(Snellman 1986)

5. Nonstandard levels

5.1. **First standard level** *(U.S. aviation forecasting).* The first standard level refers to the first thousand-foot level above the station elevation (i.e. between 0 and 1,000 ft AGL).

5.2. **Second standard level** *(U.S. aviation forecasting).* The second standard level for a reporting station is found between 1,000 and 2,000 feet above the surface, depending on the station elevation. The second standard level is used to determine low-level wind shear and frictional effects on lower atmosphere winds. To compute the second standard level, find the next thousand-foot level above the station elevation and add 1,000 feet to that level. (Federal Aviation Administration 1999)

Analysis

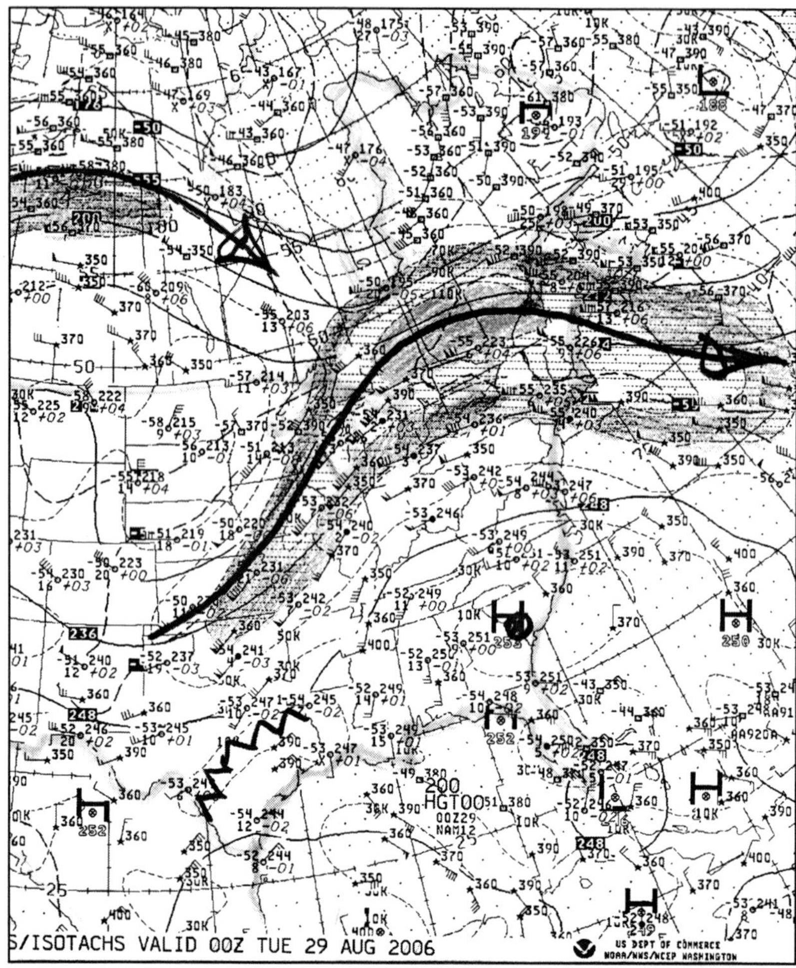

Figure B2-1. Sample of 200 mb chart for eastern North America, with hand re-analysis included. During the summer months the 200 mb chart is the highest level in the troposphere, and is valuable for locating the polar front jet and subtropical jet. Note the numerous oceanic data plots obtained from ACARS (Aircraft Communication Addressing and Reporting System), taken from airliner flights passing through the various regions shown.

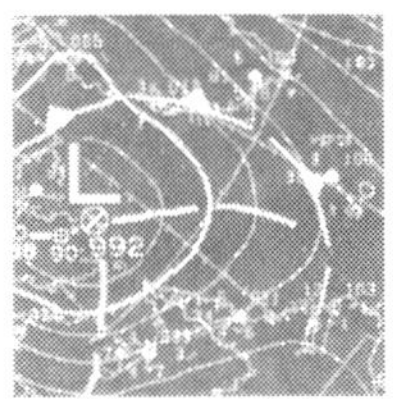

B3. ANALYSIS

Scales of motion

Weather systems are classified according to their spatial and time scales, as shown below.

Spatial scale	Timescale	Common usage	Fujita and Orlanski	Thunis & Bornstein	
		PLANETARY SCALE	MACRO-ALPHA 10000 km	MACRO-ALPHA 10000 km	General circulation Long waves
10000 km 6200 miles	weeks				
		SYNOPTIC SCALE	MACRO-BETA 2000 km	MACRO-BETA 2000 km	Extratropical cyclones Fronts
2000 km 1200 miles	2 days				
		MESO-SCALE (Sub-synoptic)	MESO-ALPHA 200 km	MACRO-GAMMA 200 km	Hurricanes Squall line
200 km 120 miles	6 hours				Mesoscale convective complexes
			MESO-BETA 20 km	MESO-BETA 20 km	Thunderstorms
20 km 12 miles	30 minutes				
			MESO-GAMMA 2 km	MESO-GAMMA 2 km	Cumulonimbus Heat island
2 km 1.2 miles	3 minutes				Clear-air turbulence
		MICRO-SCALE	MICRO-ALPHA 200 m	MESO-DELTA 200 m	
200 m 650 ft	minutes				Microbursts
			MICRO-BETA 20 m	MICRO-BETA 20 m	Tornadoes Plumes Dust devils
20 m 65 ft	1 minute				Sound waves
			MICRO-GAMMA	MICRO-GAMMA	Turbulence

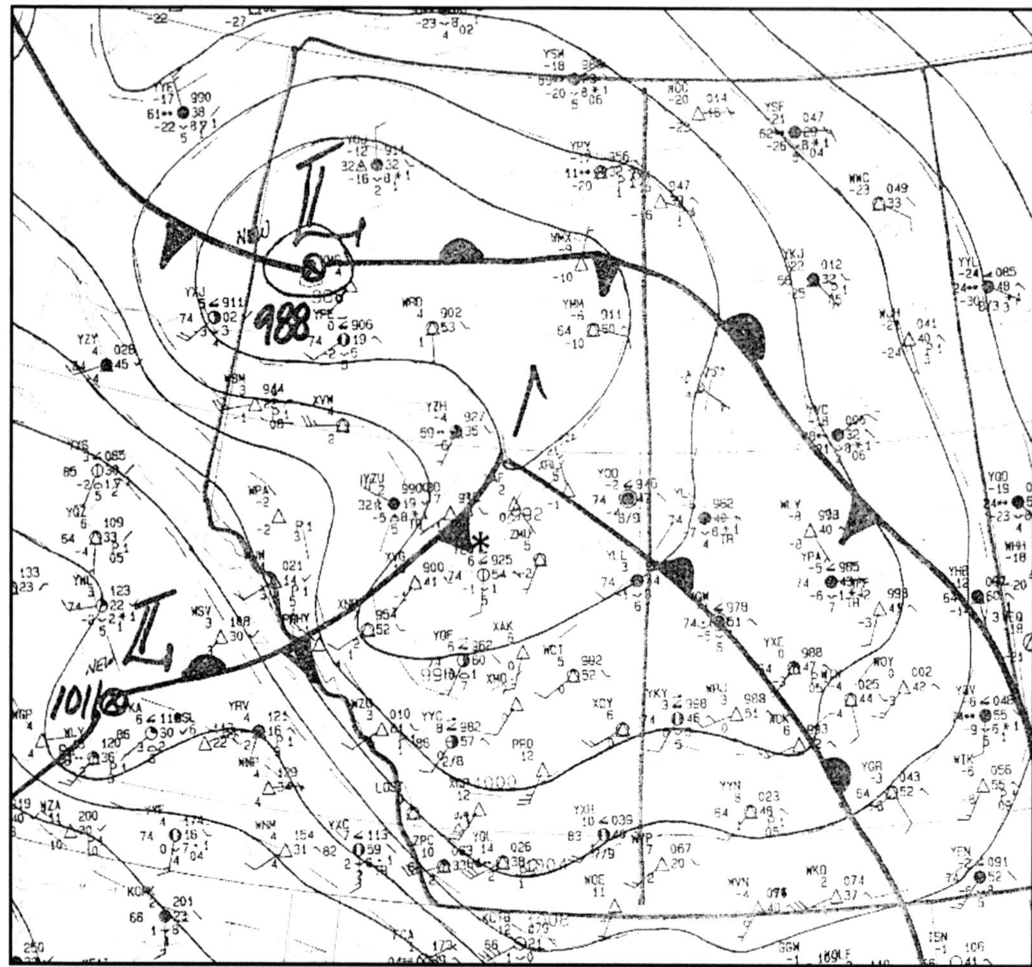

Figure B3-1. Canadian surface analysis associated with severe windstorm in the Edmonton, Alberta area (asterisk) on December 19, 2004. The analysis, covering two main provinces, is meant to bring out mesoscale detail -- a scale ideally suited to high wind events. The wind barbs behind the front contain three long shafts, signifying winds in the 30 kt (35 mph, or 56 km/h) range. The thoroughness of this hand analysis is highlighted by the hand-plotted station in western Saskatchewan between the two fronts. Chinook conditions exist in the southern Alberta area (covered in the Forecasting > Winds chapter). *(Robert Paola / Prairie and Arctic Storm Prediction Centre (PASPC), Environment Canada)*

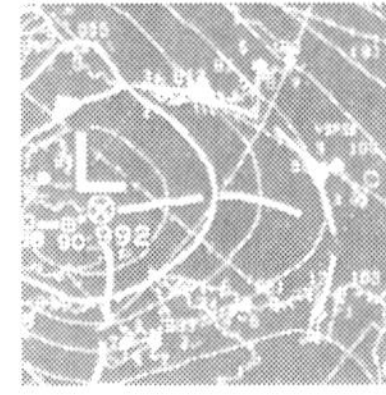

B4. ANALYSIS

Wind plots

Wind direction is always plotted with a shaft drawn away from the station center into the direction from which the wind is blowing. The shaft is always oriented relative to true north (nearly all meteorological wind directions are also coded relative to true north). A half-barb represent 5 kt, a full barb represents 10 kt, and a pennant represents 50 kt. These barbs and pennants always point outward from the shaft in a clockwise direction in the northern hemisphere, and counterclockwise in the southern hemisphere (i.e. they point geostrophically toward low pressure).

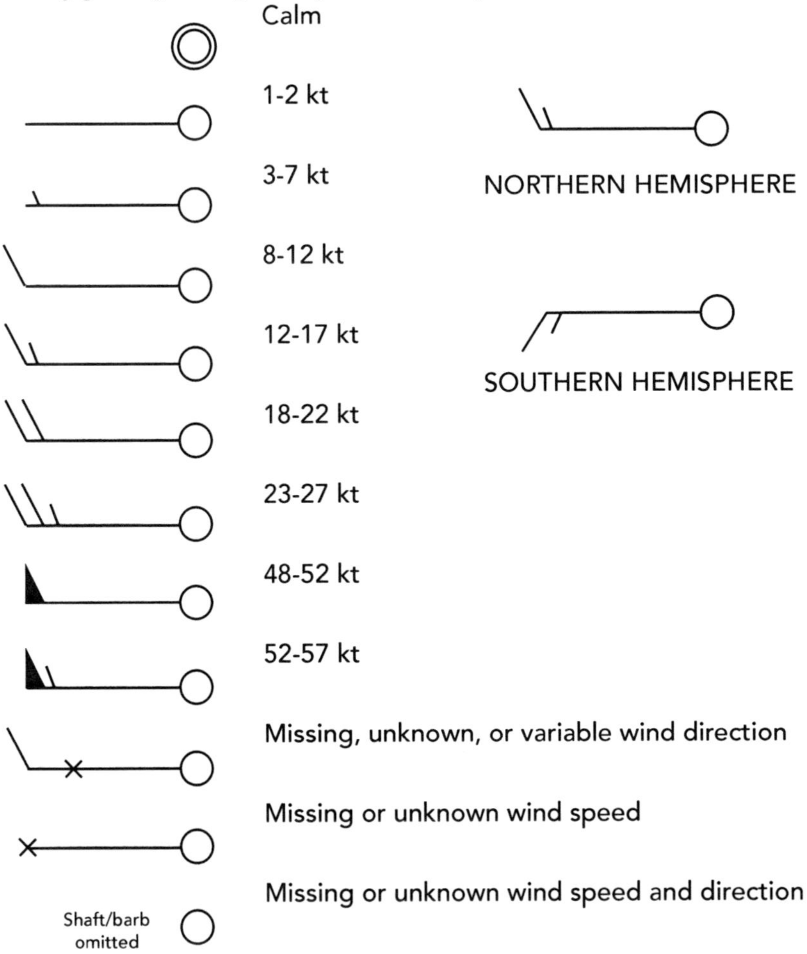

Calm

1-2 kt

3-7 kt

8-12 kt

12-17 kt

18-22 kt

23-27 kt

48-52 kt

52-57 kt

Missing, unknown, or variable wind direction

Missing or unknown wind speed

Missing or unknown wind speed and direction

Shaft/barb omitted

NORTHERN HEMISPHERE

SOUTHERN HEMISPHERE

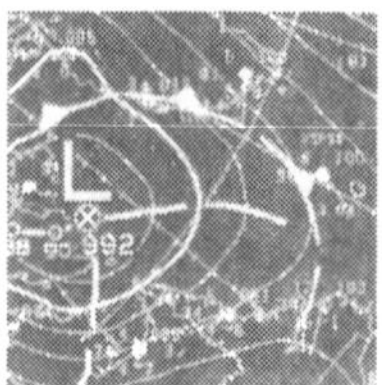

B5. ANALYSIS

Surface plots

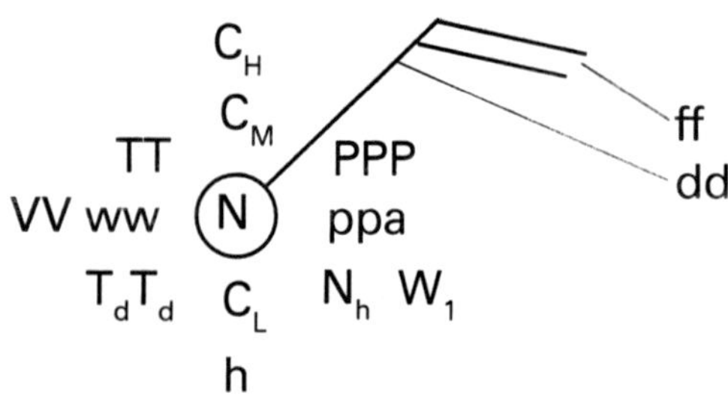

Presented here is the universally-adopted surface plot form.

TT — Temperature in degrees Celsius or Fahrenheit. Usually in whole degrees but may be expressed in tenths.

VV — Visibility in statute miles or meters. Mile values will appear as either whole numbers or fractional numbers. Meter values frequently appear as four digits, e.g. 0700.

ww — Symbol for weather type.

T_dT_d — Dewpoint temperature in degrees Celsius or Fahrenheit. Usually in whole degrees but may be expressed in tenths.

C_H — High cloud symbol.

C_M — Middle cloud symbol.

N — Total amount of cloud cover in oktas (eighths). The amount of the circle filled in is proportional to the amount of cloud cover. When an automated station produced the observation, the symbol will be plotted as a square instead of a circle. Coloring may optionally be used: blue indicates MVFR flying conditions (ceiling 1000-3000 and/or visibility 3-5 sm); and red for IFR flying conditions (ceiling less than 1000 ft and/or visibility less than 3 miles).

C_L — Low cloud symbol.

h — Height of lowest low cloud layer, or if not present, lowest middle cloud layer. This is a coded single-digit value. 0=0-50 m; 1=50-100 m; 2=100-200 m; 3=200-300 m; 4=300-600m; 5=600-1000 m; 6=1000-1500 m; 7=1500-200 m; 8=2000-2500 m; 9=2500+ m; /=unknown.

PPP — Pressure in tens, units, and tenths of a millibar. Sometimes shows units, tenths, and hundredths of an inch of mercury.

pp — 3-hour pressure change in units and tenths of a millibar.

a — A two-segmented line representing pressure change during the past three-hours.

N_h — Amount of lowest low cloud layer, or if not present, lowest middle cloud layer. Expressed in oktas (eighths); if 9 the sky is obscured.

W_1 — Symbol for type of recent weather

dd — Wind direction. Shaft points into the wind.

ff — Wind speed. A pennant represents 50 kt, a long barb represents 10 kt, and each short barb represents 5 kt. The example shows 25 kt. If the wind is calm, the shaft is omitted and a circle is drawn around the station plot.

B6. ANALYSIS

Upper air plots

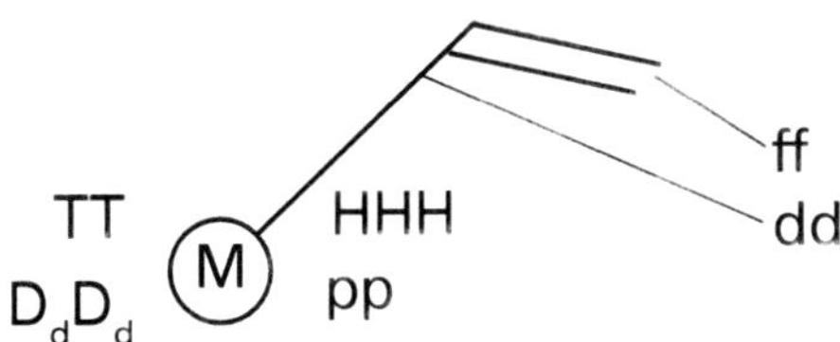

Presented here is the universally-adopted upper air plot form.

TT — Temperature in degrees Celsius. Usually in whole degrees but may be expressed in tenths.

D_dD_d — Dewpoint depression in Celsius degrees. Usually in whole degrees but may be expressed in tenths.

M — The plot circle is filled whenever the dewpoint depression is 5 Celsius degrees or less. This signifies the possible presence of cloud material and perhaps the threat for icing. The element will appear as a square when the observation was made by an aircraft (such as ACARS data) or a dropsonde. It will appear as an asterisk when the observation was satellite-based.

HHH — Geopotential height. It is either thousands, hundreds, and tens of meters or hundreds, tens, and units of meters according to the level involved. See the table at right.

pp — 12-hour height change in meters.

dd — Wind direction. Shaft points into the wind.

ff — Wind speed. A pennant represents 50 kt, a long barb represents 10 kt, and each short barb represents 5 kt. The example shows 25 kt. If the wind is calm, the shaft is omitted and a circle is drawn around the station plot.

HEIGHT DECODING RULES

1000 mb: Hundreds, tens, and units of meters. If the figure exceeds 500, subtract it from 500 to get the correct negative height.

925 mb: Hundreds, tens, and units of meters.

850 mb: Hundreds, tens, and units of meters. The thousands place is always "1".

700 mb: Hundreds, tens, and units of meters. The thousands place is always "2" or "3", whichever brings the entire figure closest to 3,000.

500, 400 mb: Thousands, hundreds, and tens of meters. In other words, it is an expression in whole decameters.

300, 250 mb: Thousands, hundreds, and tens of meters. The ten-thousands place is always "0" or "1", whichever brings the entire figure closest to 10,000.

200, 150, 100 mb: Thousands, hundreds, and tens of meters. The ten-thousands place is "1".

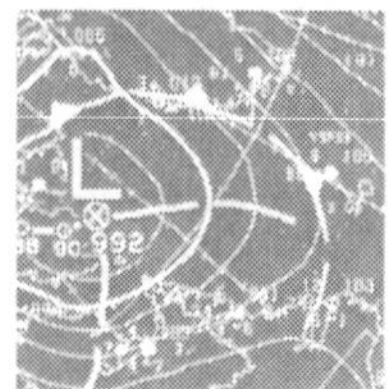

B7. ANALYSIS

Isentropic plots

Wilson plot model form

Adapted from Wilson (1985).

MMM — Montgomery streamfunction (J/kg).

FF — Wind speed (kt).

PPP — Pressure (mb)

TTT — Stability (mb between this and next higher surface)

Q — Specific humidity (g/kg)

SSS — Saturation pressure (mb)

ff — Wind speed barb

dd — Wind direction

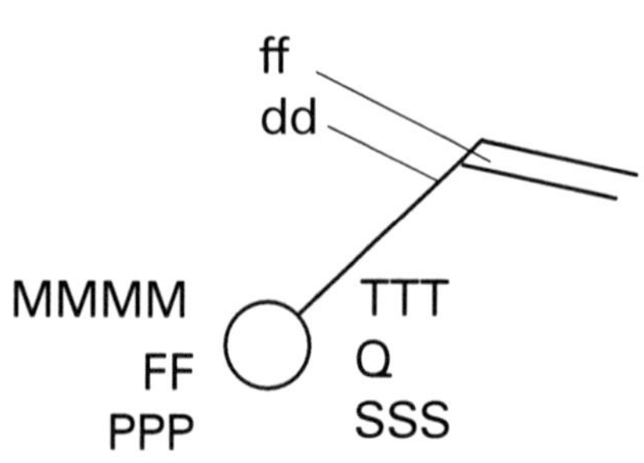

NWS AFOS plot model form

Adapted from Little (1985).

CCC — Lift required to reach condensation (mb)

RRR — Mixing ratio at isentropic surface (g/kg/10)

PPP — Pressure (mb)

ff — Wind speed barb

dd — Wind direction

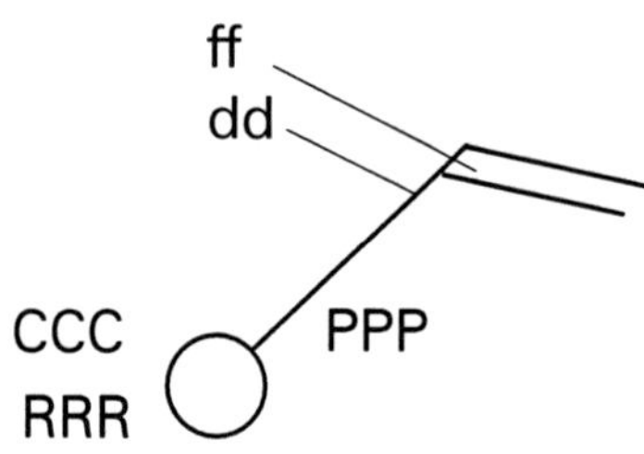

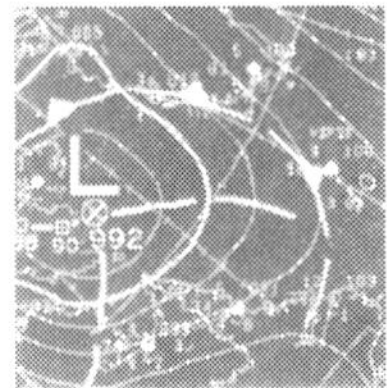

B8. ANALYSIS

SKEW-T log p

Listed below are the components of a SKEW-T log P diagram.

1. Temperature

Temperatures are diagonal brown lines which slope from top right to bottom left. The lines are used to represent either temperature or dewpoint (the adiabats do not apply to plotted dewpoint temperatures).

2. Pressure

Pressure lines are horizontal brown lines, calibrated in millibars. They are inversely proportional to height.

3. Dry adiabat

The dry adiabat is represented by a solid brown line sloping from top left to bottom right. It represents the change in temperature of a *dry parcel as it rises or sinks*. It is equivalent to potential temperature.

4. Moist adiabat

Also known as ***wet adiabat; saturation adiabat***

The moist adiabat is represented by solid, curved green lines that stand almost vertically. They represent the change in temperature of a *saturated parcel as it rises*. They can also be used to determine wet-bulb temperature by lifting the parcel along the dry adiabat to its lifted condensation level, then extrapolating a line down the moist adiabat back to the parcel's original level.

5. Mixing ratio

Mixing ratio lines are represented by straight, dashed green lines that stand almost vertically. When used for a plotted parcel temperature, they represent its saturation mixing ratio. When used for a plotted parcel dewpoint, they represent its mixing ratio.

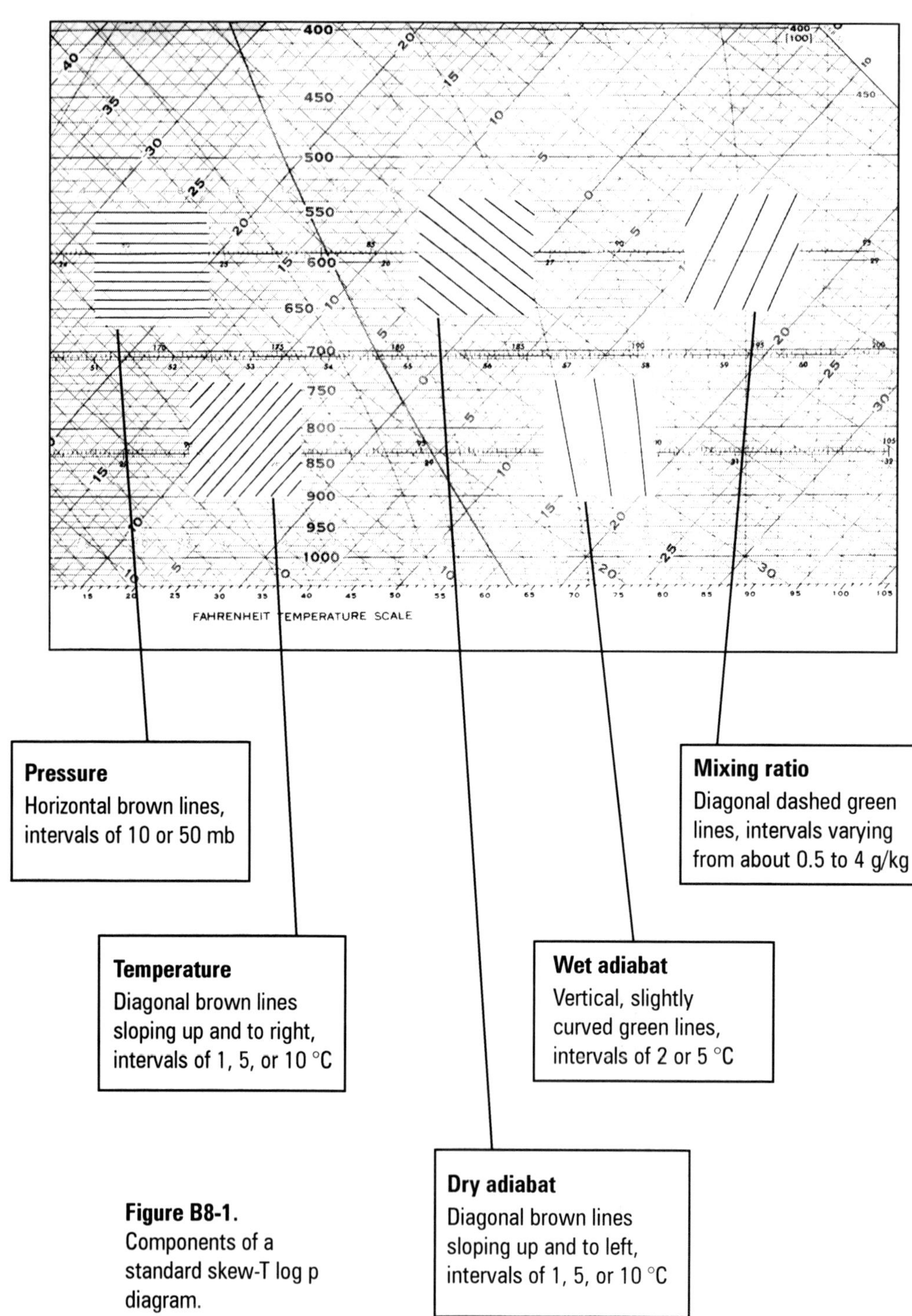

Figure B8-1. Components of a standard skew-T log p diagram.

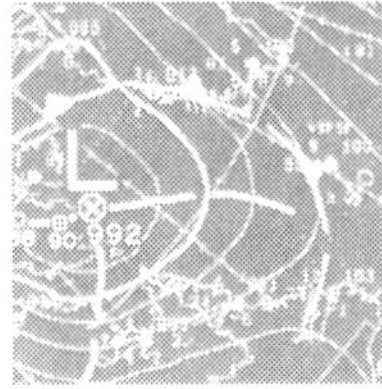

B9. CHART ANALYSIS

Analysis markings

L

Low pressure center

H

High pressure center

Tropical storm Northern Hemisphere

Hurricane, typhoon, or tropical cyclone Northern Hemisphere

Tropical storm Southern Hemisphere

Hurricane, typhoon, or tropical cyclone Southern Hemisphere

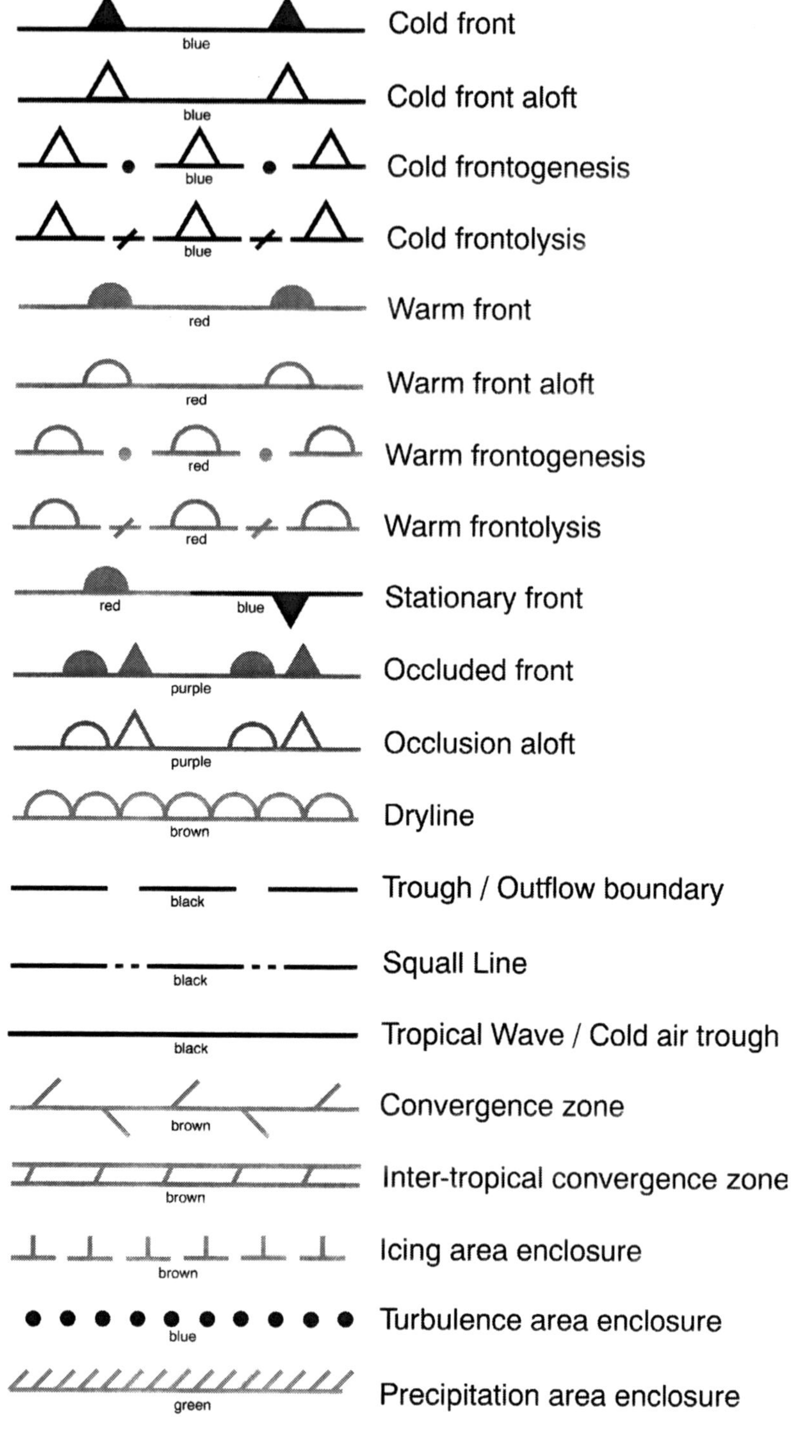
blue
Cold front
blue
Cold front aloft
blue
Cold frontogenesis
blue
Cold frontolysis
red
Warm front
red
Warm front aloft
red
Warm frontogenesis
red
Warm frontolysis
red
blue
Stationary front
purple
Occluded front
purple
Occlusion aloft
brown
Dryline
black
Trough / Outflow boundary
black
Squall Line
black
Tropical Wave / Cold air trough
brown
Convergence zone
brown
Inter-tropical convergence zone
brown
Icing area enclosure
blue
Turbulence area enclosure
green
Precipitation area enclosure

Beginning from this point forward are standard markings used by Air Force Global Weather Central during the 1950s and 1960s, as published by Col. Robert C. Miller in *Notes on Analysis and Severe-Storm Forecasting Procedures of the Air Force Global Weather Central* (1972). Miller created one of the few sets of standardized meteorological symbols, illustrated here in full.

	COLOR	MONOCHROME	
varies			Height change isopleth
black			Thickness ridge
black			Thickness no-change line
black			Thickness fall isopleth
black			Wet-bulb zero isopleth
black			Anticyclonic shear
black			Level of free convection
black			Vertical Totals (VT) Index isopleth
black			Cross Totals (CT) Index isopleth
orange			Total Totals (TT) Index isopleth
black			Lifted Index (LI) isopleths
blue			Outer severe weather area
red			Primary severe weather area

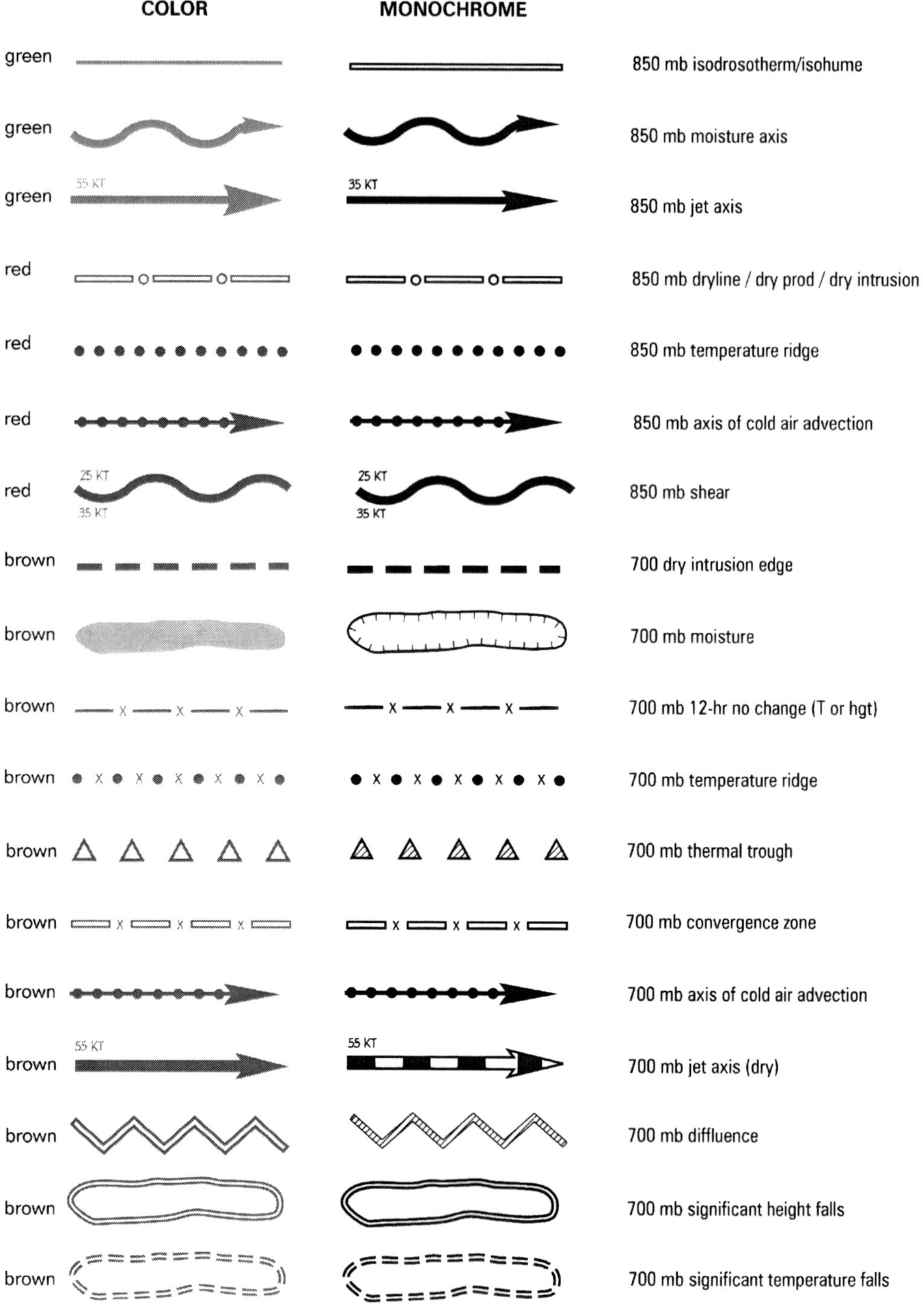
COLOR
MONOCHROME
green
850 mb isodrosotherm/isohume
green
850 mb moisture axis
green
55 KT
35 KT
850 mb jet axis
red
850 mb dryline / dry prod / dry intrusion edge
red
850 mb temperature ridge
red
850 mb axis of cold air advection
red
25 KT
35 KT
25 KT
35 KT
850 mb shear
brown
700 dry intrusion edge
brown
700 mb moisture
brown
700 mb 12-hr no change (T or hgt)
brown
700 mb temperature ridge
brown
700 mb thermal trough
brown
700 mb convergence zone
brown
700 mb axis of cold air advection
brown
55 KT
55 KT
700 mb jet axis (dry)
brown
700 mb diffluence
brown
700 mb significant height falls
brown
700 mb significant temperature falls

	COLOR	MONOCHROME	
lue			500 mb isotherms
lue			500 mb critical isotherm
lue			500 mb thermal trough
ue	-16	-16	500 mb significant height falls
lue	-4	-4	500 mb significant temperature falls
reen			500 mb moisture
ellow			500 mb PVA zone
lue	70 KT	70 KT	500 mb jet axis
lue			500 mb shear
lue			500 mb diffluence
urple	90 KT	90 KT	300-200 mb jet axis
urple			300-200 mb jet max
urple			300-200 mb diffluence
urple			300-200 mb shear

Analysis

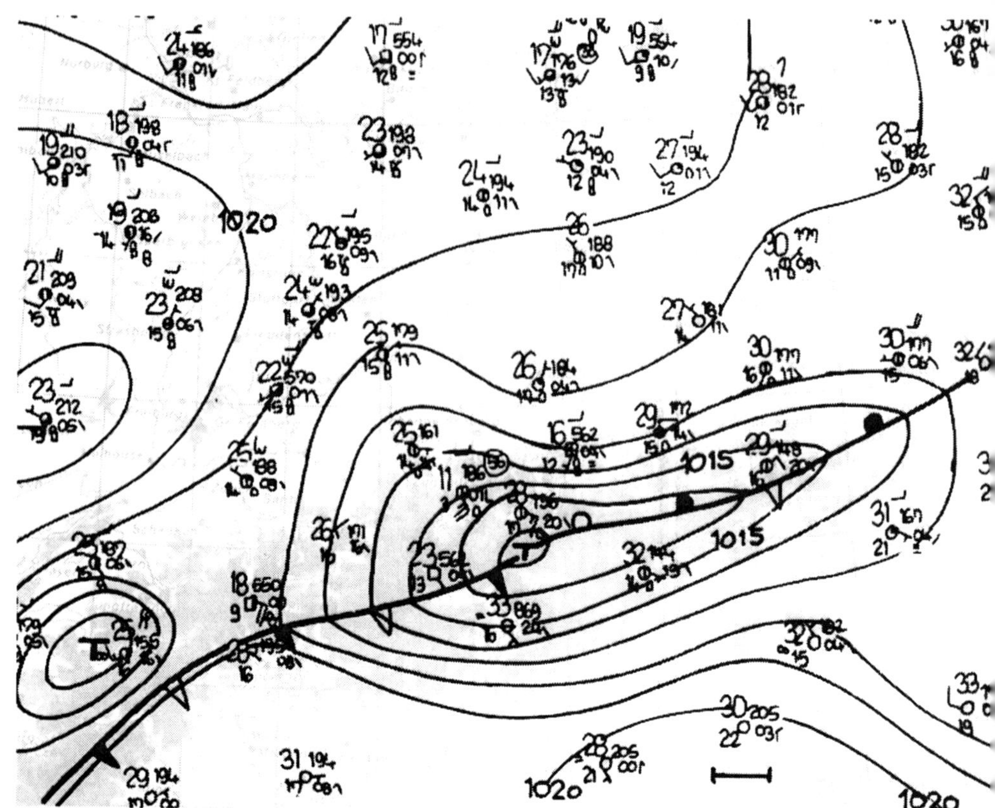

Figure B9-1. German Deutscher Wetterdienst surface weather chart for 12 July 1984. Though a language barrier exists, meteorologists from many different cultures can readily analyze and interpret one another's charts thanks to common standards adopted globally. *(Deutscher Wetterdienst)*

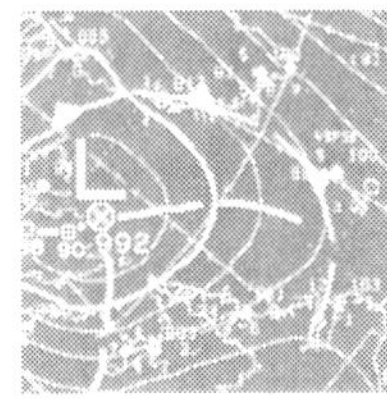

B10. ANALYSIS

Color standards

1. Meteorological depiction

These color schemes are intended for the use of forecasters.

1.1. **Frontal analysis**.

Element	Color
Cold front	Blue
Warm front	Red
Stationary front	Alternating blue and red
Occluded front	Purple
Dryline	Brown
Trough	Black or brown

1.2. **Upper-level analysis**. The dewpoint depression scheme is adapted from (CMC 1987).

Element	Color
Short wave trough	Solid red line (CMC)
Short wave ridge	Solid blue line (CMC)
Dewpoint depression ≤2°C	Solid green line enclosing green shading
Dewpoint depression 2-5°C	Solid green line enclosing green hatching
Dewpoint depression >20°C	Solid brown line enclosing brown shading

1.3. **Streamline analysis**.

Element	Color
Confluent asymptotes	Red
Difluent asymptotes	Blue

1.4. **Wind**. The following scheme is a 1987 scheme for isotach shading proposed by the Canadian Meteorological Center.

Range	Shading
60-90 kt	Green
90-120 kt	Red
120-150 kt	Blue
150-180 kt	Purple
180+ kt	Yellow

(Canadian Meteorological Center 1987)

1.5. **Clouds**. The following scheme is used by U.S. Air Force forecasters:

Ceiling	Outline color
Less than 1000 ft	Solid red line with shaded red fill
1000 to 2900 ft	Scalloped blue line with inner edge shaded blue
3000 to 9500 ft	Scalloped purple line with inner edge shaded purple
10000 ft or more	Scalloped brown line with inner edge shaded brown

1.6. **Icing**. The following scheme is one used by U.S. Air Force forecasters which is based on icing height:

Icing height	Color
5000 ft or less	Red
10,000 ft or less	Blue
15,000 ft or less	Green

2. Public depiction

The following elements list color standards recommended for the public sector (American Meteorological Society, 1993) with exceptions as noted. Meteorological analysis conventions for internal use may differ and are covered in a separate section.

2.1. **Linear features**.

- Cold front: Blue.
- Warm front: Red.

- Occluded front: Purple.
- Stationary front: Alternating blue and red.
- Trough / outflow: Brown. However, black is in common use.
- Dryline: Brown.
- Jet stream: Light blue.
- Contours: Tan.

2.2. **Typographic features**.

- Thunderstorm symbol: Red.
- Tropical storm and hurricane symbol: Red.
- Rain / rain shower symbol: Green.
- Snow / snow shower symbol: White.
- High pressure symbol: Black. Blue is in common use.
- Low pressure symbol: Black. Red is in common use.

2.3. **Areal features**.

- Rain areas: Green.
- Snow areas: White.
- Mixed precipitation areas: Green. Orange common.
- Fog areas: Yellow.
- Blowing dust areas: Brown.
- Turbulence areas: Yellow/orange/red.
- Icing areas: Yellow/orange/red.
- Moisture aloft: Green.
- Marginal visual flight rules (MVFR) areas: Yellow.
- Instrument flight rules (IFR) areas: Red.
- High wind areas: Blue.
- Severe weather areas: Red.
- Air quality areas: Brown.
- Fire danger areas: Yellow/orange/red.
- High seas areas: Yellow/orange/red.

- Convective outlook areas: Yellow/orange/red.
- Tropical storm areas: Orange.
- Hurricane areas: Red.
- Freeze areas: Purple.
- Heavy snow areas: Gray.
- Blizzard areas: White.
- Small craft advisory areas: Yellow.
- Gale areas: Orange.
- Maritime storm areas: Red/orange.

Figure B10-1. Color definitions in RGB decimal and 8-bit hexadecimal.

Color	RGB decimal	RGB hex
White	255,255,255	$FFFFFF
Gray	192,192,192	$C0C0C0
Black	000,000,000	$000000
Red	255,000,000	$FF0000
Light red	255,128,128	$FF8080
Reddish orange	255,064,000	$FF4000
Orange	255,128,000	$FF8000
Yellow	255,255,000	$FFFF00
Tan	192,128,000	$C08000
Brown	128,064,000	$804000
Green	000,255,000	$00FF00
Light green	128,255,128	$80FF80
Blue	000,000,255	$0000FF
Light blue	000,128,255	$0080FF
Purple	128,000,255	$8000FF

SECTION C

Forecasting

AREA FORECAST DISCUSSION...UPDATED
NATIONAL WEATHER SERVICE NEW ORLEANS LA
452 PM CDT SUN AUG 28 2005

..UPDATE...TO ADD TORNADO WATCH #752.

..DISCUSSION...
SOUTHEAST LOUISIANA SEEMS POISED FOR A DATE WITH DESTINY AS CATEGORY 5 HURRICANE KATRINA CONTINUES TO KEEP A BEAD ON BARATARIA BAY AND THE GREATER NEW ORLEANS AREA. THE GFS MODEL CONTINUES TO BE SUPERIOR IN ITS HANDLING OF THE SYSTEM INASMUCH AS TO BASE THE CONVENTIONAL FORECAST PARAMETERS WITH GOOD INTEGRITY AND IN AGREEMENT WITH NHC ADVISORIES.

NEEDLESS TO SAY...THE WORST CAN BE ANTICIPATED AND URGENCY IS BEING STRESSED IN ALL PRODUCTS AS A WORST CASE HURRICANE SCENARIO FOR THIS VERY FRAGILE AND VULNERABLE STRETCH OF U.S. COASTLINE. THE EYE IS EMERGING ON THE KLIX LONG RANGE LOOP AND BANDS ARE EXTENDING TO LAKE PONTCHARTRAIN AT THIS TIME. THINGS WILL BE DETIORATING STEADILY FROM THIS POINT FORWARD FOR THE NEXT 24 HOURS.

WILL MAINTAIN ALL WARNINGS AS ALREADY POSTED AS WELL AS THE FLASH FLOOD WATCH. STORM PREDICTION CENTER HAS ADVISED THAT THE FIRST TORNADO WATCH OF THE EVENT WILL LIKELY BE ISSUED FOR THE REGION EARLY THIS EVENING...PROBABLY RIGHT AFTER SUNSET.

MOST ATTENTION WITH THIS PACKAGE WAS DAY 1-2 WITH LITTLE IF ANY CHANGES MADE BEYOND DAY 3. GOOD LUCK AND GODSPEED TO ALL IN THE PATH OF THIS STORM.

NNNN

URGENT - WEATHER MESSAGE
NATIONAL WEATHER SERVICE NEW ORLEANS LA
619 AM CDT MON AUG 29 2005

EXTREMELY DANGEROUS HURRICANE KATRINA MOVING ACROSS LOWER PLAQUEMINES PARISH

DEVASTATING DAMAGE EXPECTED

THE MAJORITY OF INDUSTRIAL BUILDINGS WILL BECOME NON FUNCTIONAL. PARTIAL TO COMPLETE WALL AND ROOF FAILURE IS EXPECTED. ALL WOOD FRAMED LOW RISING APARTMENT BUILDINGS WILL BE DESTROYED. CONCRETE BLOCK LOW RISE APARTMENTS WILL SUSTAIN MAJOR DAMAGE...INCLUDING SOME WALL AND ROOF FAILURE.

HIGH RISE OFFICE AND APARTMENT BUILDINGS WILL SWAY DANGEROUSLY...A FEW POSSIBLY TO THE POINT OF TOTAL COLLAPSE. MANY WINDOWS WILL BLOW OUT.

AIRBORNE DEBRIS WILL BE WIDESPREAD...AND MAY INCLUDE HEAVY ITEMS SUCH AS HOUSEHOLD APPLIANCES AND EVEN LIGHT VEHICLES. SPORT UTILITY VEHICLES AND LIGHT TRUCKS WILL BE MOVED. THE BLOWN DEBRIS WILL CREATE ADDITIONAL DESTRUCTION. PERSONS...PETS...AND LIVESTOCK EXPOSED TO THE

It is not the answer that enlightens, but the question.

EUGENE IONESCO, 1969
Romanian-French playwright

C1. FORECASTING

Methodology

Weather forecasting encompasses several primary techniques:

1. Climatology

Climatology establishes a baseline for expected weather, based on mean observed values for a specific station and a specific date or range of dates.

"Fog is expected on 21 out of 30 nights this month."

2. Analogs

Analog forecasting refers to the use of a nearly-identical weather pattern on a past date. The observed weather that followed the previous weather pattern establishes a baseline for the forecast.

"This type of weather pattern two weeks ago brought fog."

3. Persistence

Persistence is an assumption that weather patterns which are occurring now will continue to occur. It works best in stagnant weather patterns.

"Fog occurred last night, so fog will occur tonight."

4. Trends/extrapolation

Trend forecasting is the extrapolation of changes which are occurring now into the future.

"The nights have brought successively more widespread fog."

5. Numerical weather prediction

Numerical weather prediction is a form of quantitative forecasting that uses equation-solving on a model of the actual atmosphere. The solutions of these equations determine the forecast weather.

"The WRF is forecasting fog for this area."

6. Ensemble forecasting

Ensemble forecasting uses a combination of numerical predictions to forecast the weather.

"The ensemble model output indicates fog will occur."

7. Empirical prediction

Empirically-based forecasting, which includes rules of thumb, lore, and deductive reasoning, makes use of forecasting rules which are constructed from observation or experience and show strong forecast skill. They may or may not have scientific basis, and in some cases the scientific basis may be poorly understood.

"Given our observed wind and sky cover, we usually get fog."

C2. FORECASTING

Surface systems

1. Baroclinic low

1.1. **Characteristics**. Low pressure with thermal advection. Stacks towards the cold air with height.

1.2. **Manifestations**. Frontal low; extratropical low; stable wave.

1.3. **Forecasting notes on movement**.

- Baroclinic lows that are not occluded will move parallel to the warm sector isobars. [ATC/ER]
- Dynamic lows tend to move at 70% of the 700 mb winds and 50% of the 500 mb winds. [ATC/ER]
- A low should move rapidly if the low level warm advection is strong and in phase with the divergence aloft. [ATC/ER]
- Lows moving southeast usually intensify and slow down as they begin to recurve to the northeast. [ATC/ER]
- **Rule of persistence**: Lows will tend to follow a track similar to previous systems until the long wave pattern changes. [ATC/ER]
- When the thickness gradient and the mean 700 mb wind flow are equal, a low will move in a direction midway between the two. Otherwise it will move in the direction of the stronger flow. [ATC/ER]
- A low which has a warm front extending to the southeast and a cold front extending to the west or northwest will move southeast parallel to the thickness lines along the warm front. Once the cold front has reached the southwest quadrant of the low and the warm front is in the southeast quadrant, the system will eventually occlude. [ATC/ER]

- Stable waves on a frontal boundary will move along the edge of a cold air mass parallel to the 1000-500 mb thickness lines, and are usually short-lived. [ATC/ER]

1.4. **Forecasting notes on intensity**.

- Frontal waves tend to form on stationary or on slow-moving cold fronts. [ATC/ER]
- Surface lows that move to the right (left) of the normal track will move toward higher (lower) contour heights and tend to fill (deepen). [ATC/ER]
- A surface low will usually not deepen more than 1 mb per hour, except in typical "bomb" formation areas.
- Surface lows typically move parallel to the 500 mb jet.
- A surface low will deepen at about 8 mb for every 60 m of 500 mb height falls.
- When an extratropical cyclone is embedded in *diffluent, high-amplitude flow*, the structure resembles the Norwegian cyclone model. When the cyclone is embedded in *confluent, low-amplitude flow*, the structure resembles the Shapiro-Keyser model. (Schultz, 1998)
- **Rosenbloom Rule**: Rapidly deepening surface lows tend to move to the left of the numerically forecast track.

2. Baroclinic high

2.1. **Characteristics**. High pressure with thermal advection. Stacks towards the warm air with height.

2.2. **Manifestations**. Frontal high.

2.3. **Forecasting notes**.

- Surface highs will usually build if they are under 500 mb height rises.
- Surface highs will generally move equatorward even when the upper flow is westerly. [ATC/ER]

- Surface highs tend to pass around the Great Lakes in the late fall and early winter.
- Cold highs will move rapidly southeast once a northwesterly jet moves over them. [ATC/ER]
- Once a dynamic high starts changing into a warm [barotropic] high, its movement will slow down. [ATC/ER]

3. Cold-core barotropic low

Technically "equivalent barotropic".

3.1. **Characteristics**. Low pressure at the surface becoming more intense with height. Stacks vertically. They are most common during the winter season.

3.2. **Manifestations**. The two primary types of cold-core barotropic lows are the decaying wave (poleward of the polar front jet) and the cut-off low (equatorward of the polar front jet).

3.3. **Forecasting notes**.

- Once an unstable wave begins to occlude, it will tend to recurve in a more northerly direction.
- Occluded lows will move in a direction parallel to the isobars ahead of the warm front. [ATC/1]

4. Warm-core barotropic low

Technically "equivalent barotropic".

4.1. **Characteristics**. Low pressure at the surface weakening with height and becoming high pressure aloft. With tropical cyclones the high aloft is generally seen only in the uppermost troposphere. Stacks vertically.

4.2. **Manifestations**. Warm-core barotropic lows caused by intense surface heating and with little latent heat release are known as thermal ("heat") lows. Warm-core barotropic lows that develop over warm waters and involve considerable latent heat release are known as tropical cyclones (hurricanes, typhoons, and tropical storms).

4.3. **Forecasting notes**.

- See *Tropical Cyclones.*

5. Cold-core barotropic high

Technically "equivalent barotropic".

5.1. **Characteristics**. High pressure at the surface weakening and becoming weak high pressure or low pressure aloft. Stacks vertically. It is generally caused by rapid, intense cooling in the lower troposphere. The upper reflection is often not well-defined.

5.2. **Manifestations**. Polar air in source region; plateau high.

5.3. **Forecasting notes**. None.

6. Warm-core barotropic high

Technically "equivalent barotropic".

6.1. **Characteristics**. High pressure at the surface becoming more intense with height. Stacks vertically. The surface reflection is often not well-defined.

6.2. **Manifestations**. Subtropical highs (e.g. the "Bermuda High"); cutoff highs.

6.3. **Forecasting notes**.

- Once a dynamic high starts changing into a warm [barotropic] high, its movement will slow down. [ATC/ER]
- Warm highs will move with the speed of the upper-level ridge. [ATC/ER]

7. Special surface patterns

7.1. **Cold air damming**. Cold air damming is favored when statically stable air is driven into a perpendicularly-oriented barrier (a mountain range) by a surface cyclone equatorward or an anticyclone poleward of a forecast area. (Dunn 1987)

- The north-south mountains in the United States provide excellent configurations for damming. Cold air damming is common east of the Appalachians, especially in the Carolinas, during the winter. It may occur on a larger scale east of the Rockies.
- The cold air mass is usually shallow and poorly represented by 1000-500 mb values. Winds in the cold air mass are highly ageostrophic and tend to be northerly (in U.S.) despite the pressure gradient.
- If southwesterly flow above the surface is present, "overrunning" (isentropic lift) may produce precipitation episodes.

7. Regional techniques

7.1. **Western United States**.

- A Rocky Mountain lee-side trough will typically begin developing when the 500 mb ridge axis reaches within 50 to 300 miles west of the lee slopes. The surface low remains stationary and deepens until the ridge axis is 50 to 300 miles east of the slopes. The surface low moves out at the speed of the upstream 500 mb trough.
- In Nevada during the winter, numerical models, particularly the old LFM, have historically underforecast low development with a digging 500 mb trough approaching the Pacific coast.
- A plateau high over the Great Basin will be stationary without a thickness gradient, or will drift parallel to the contours. It will break down quickly when fast zonal westerlies develop overhead.
- California inverted surface heat troughs usually develop within a building 500 mb ridge. The trough axis lies under that part of the 500 mb ridge south of the 500 mb jet and is most pronounced where the 500 mb winds are less than 25 kt. During the warm season, this trough may extend into Oregon and Washington.

7.2. **Central United States**.

■ Maritime polar high centers will not usually move east across the Rockies behind a Pacific cold front. Instead a new high center or ridge will form over the Plains states following frontal passage. [ATC/ER]

■ Arctic outbreaks will generally not reach Texas unless an upper ridge is over or forecast to be over southwestern Canada. This rule does not typically apply in January or February, but the cooling is not so great. [WSFO/FTW]

7.3. **Eastern United States**.

■ Lows tend to slow down and intensify while passing over the Great Lakes in the fall and winter. Highs will slow down and build while passing over the Great Lakes during the spring and summer. [ATC/ER]

GALLERY OF WEATHER SYSTEM TYPES

Baroclinic low

EXTRATROPICAL CYCLONE

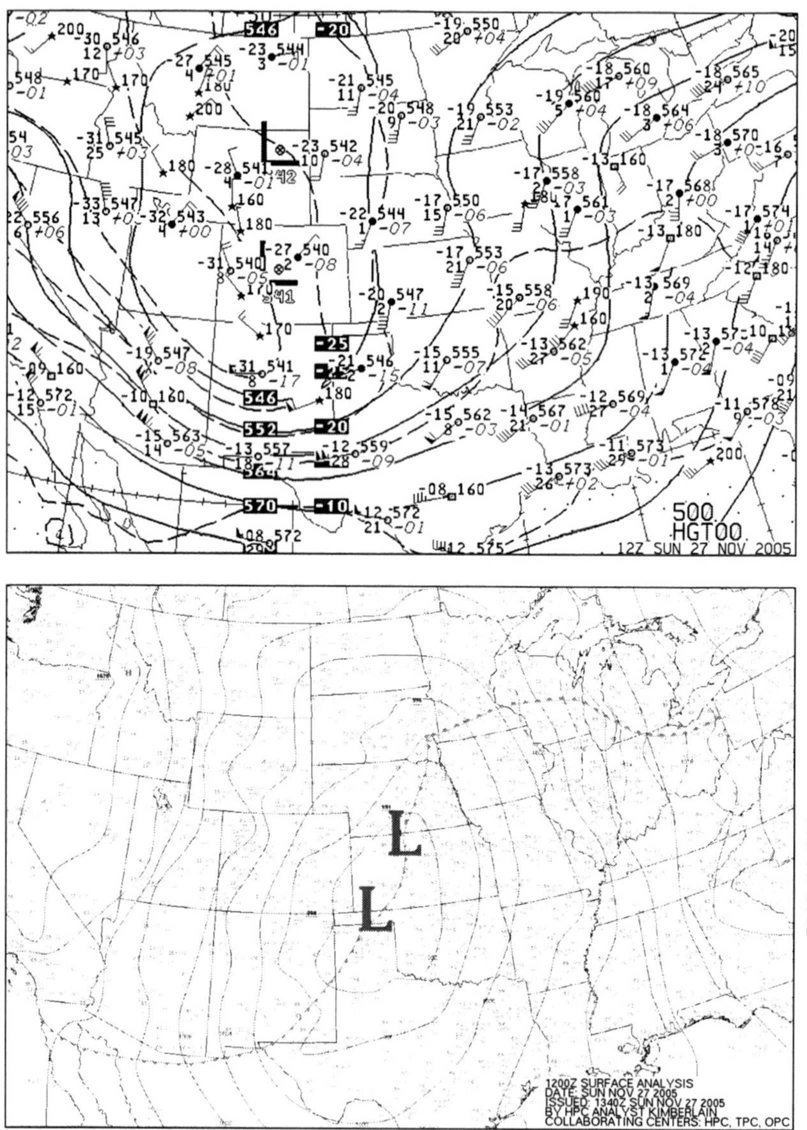

Figure C2-1a. Baroclinic low. Usually associated with the developing and mature phase of extratropical cyclones. The jet stream tends to cross the surface low. The surface low stacks westward or poleward with height towards the coldest pool of low- and mid-level.

Baroclinic high
MIGRATORY POLAR AIR MASS

500 MB

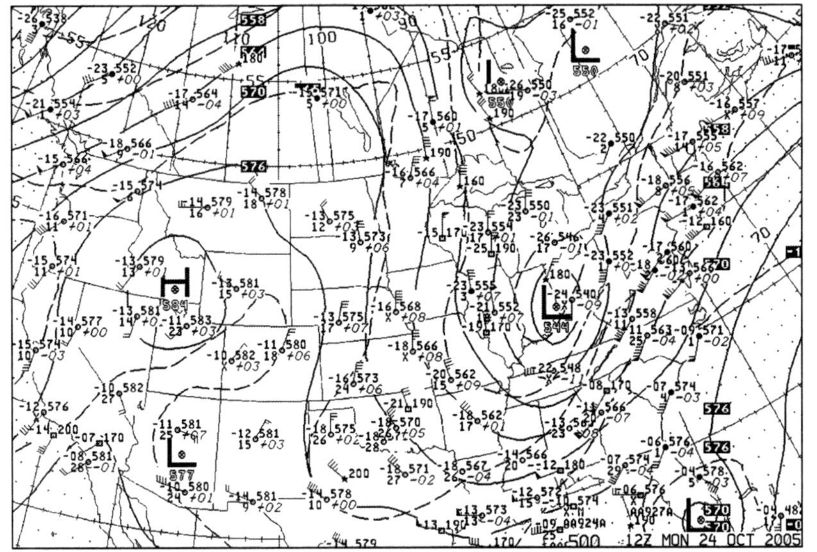

SURFACE

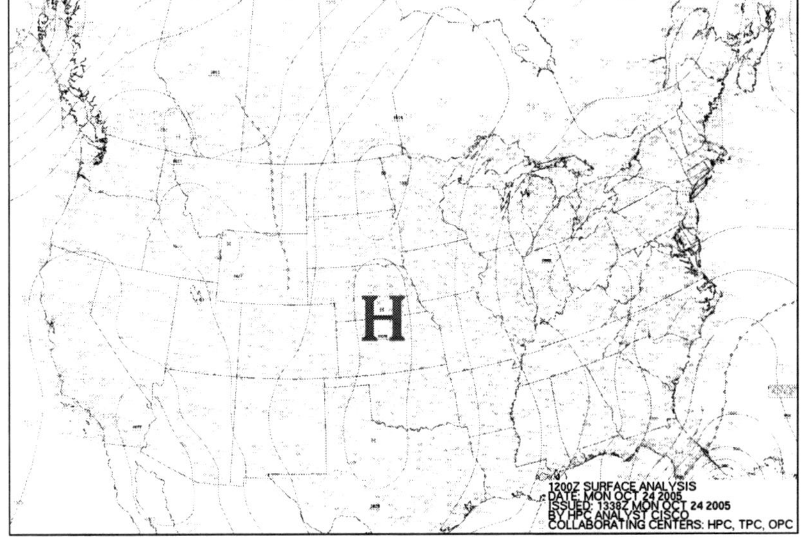

Figure C2-1b. Baroclinic high. Usually associated with polar air masses that feed into extratropical cyclones. The upper-level jet is located over the surface high. The high usually stacks westward or southwarrd with height towards the warmest low- and mid-level air.

Cold-core barotropic low

DECAYING WAVE

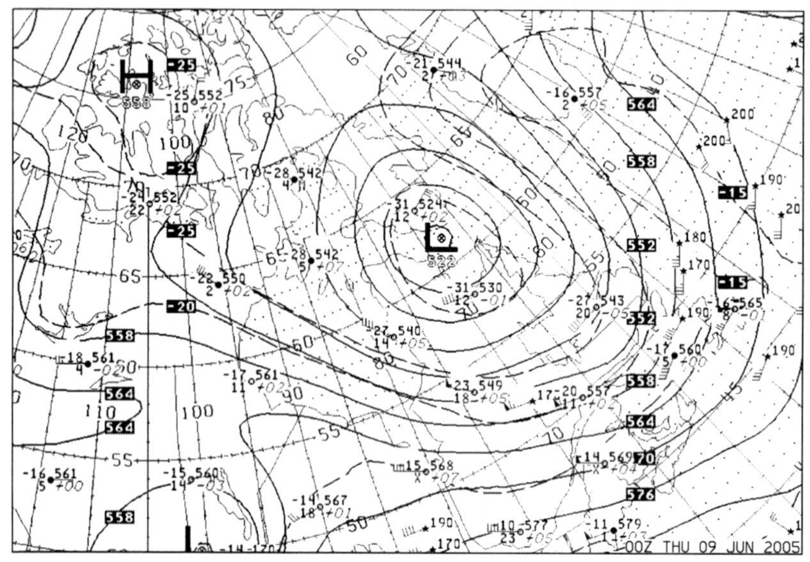

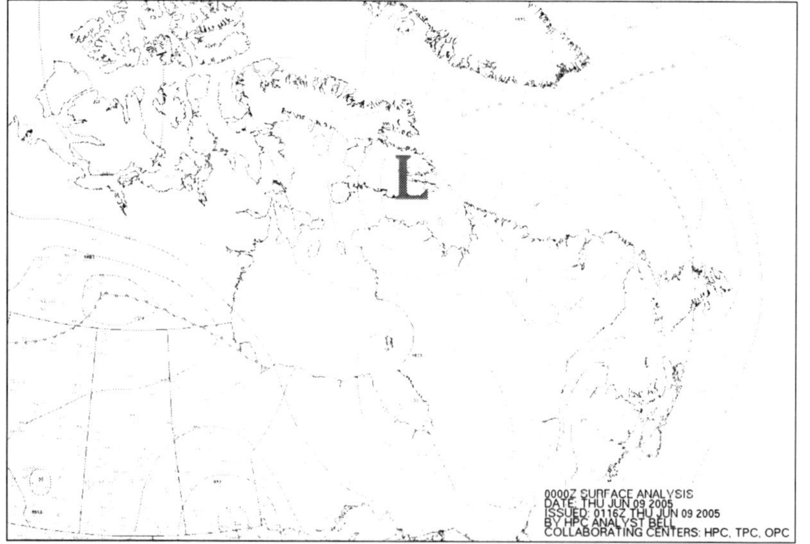

Figure C2-1c. Cold-core barotropic low (decaying wave). Associated with the decaying phase of an extratropical low. The surface low stacks vertically with height and intensifies with height. It is most common near Iceland, in the Aleutians, and in Hudson Bay. The low is deep and slow-moving, sometimes moving westward. It is most common in the winter.

Cold-core barotropic low

CUTOFF LOW

500 MB

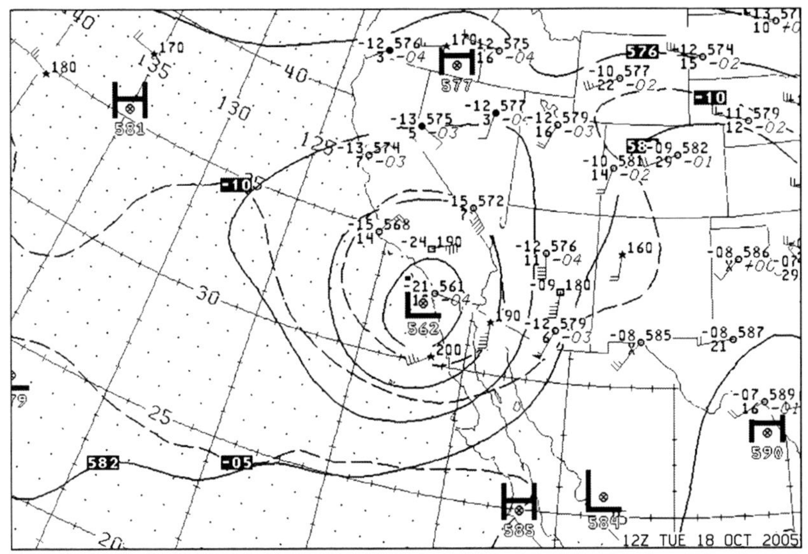

SURFACE

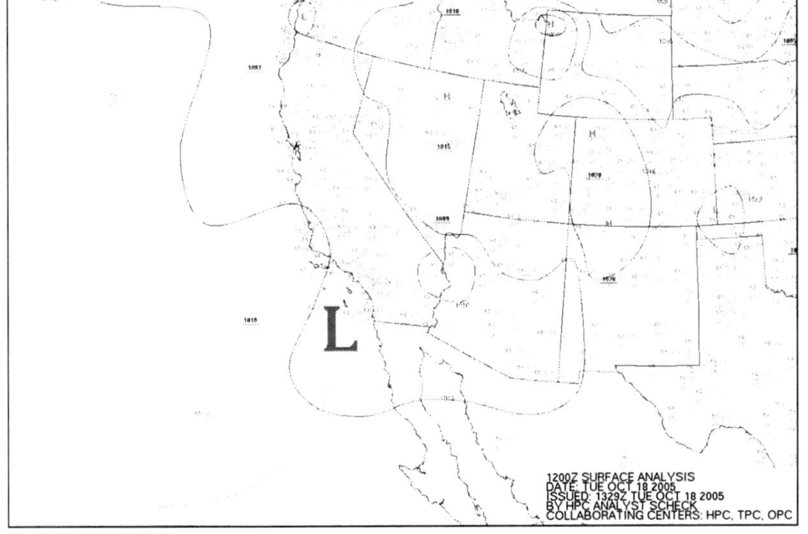

Figure C2-1d. Cold-core barotropic low (cutoff low). Associated with an upper-level low which has detached equatorward from the main band of westerlies. It is occasionally seen in the southwestern United States during the cool season. There is usually poor surface reflection but considerable precipitation fields. The circulation intensifies with height.

Warm-core barotropic low

THERMAL LOW

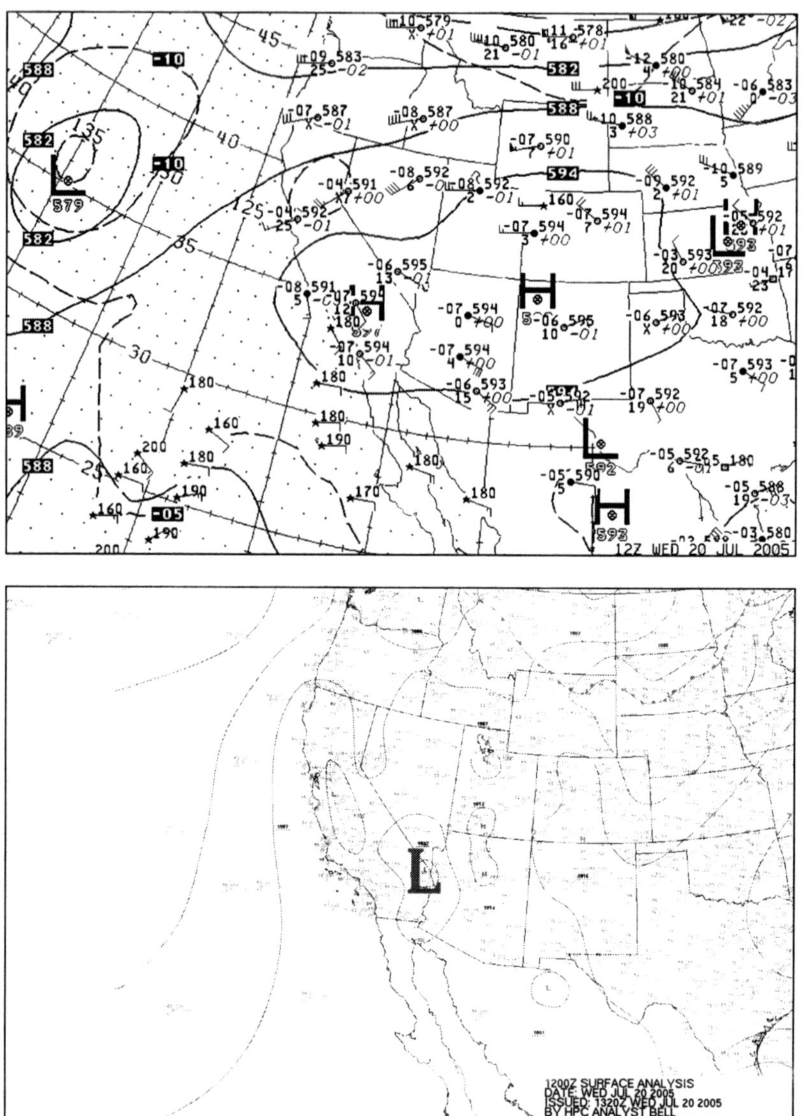

Figure C2-1e. Warm-core barotropic low (thermal low). The heat low involves no release of latent heat and is formed from very strong surface heating. It diminishes with height and becomes a high pressure area aloft. It is a semipermanent feature in the southwest U.S. and Mexico during the summer.

Warm-core barotropic low
TROPICAL CYCLONE

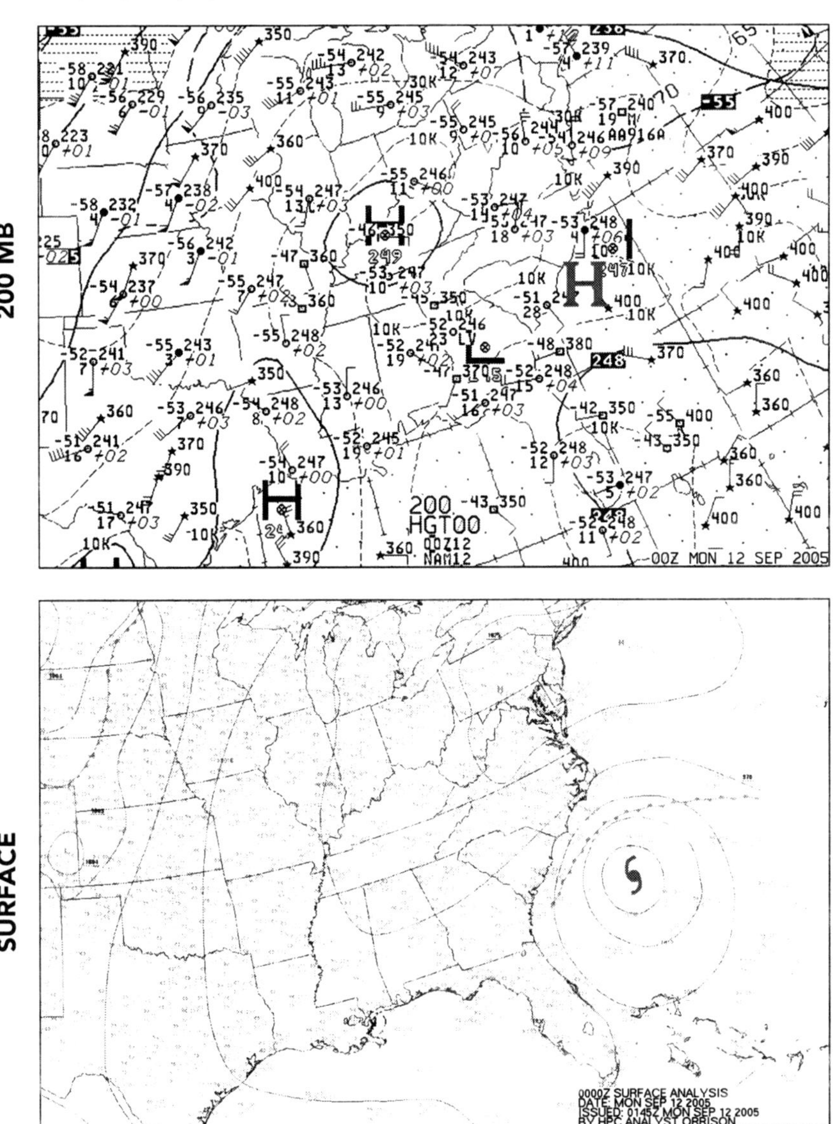

Figure C2-1f. Warm-core barotropic low (tropical cyclone). The tropical cyclone involves considerable release of latent heat. It is a cyclostrophically-balanced system that diminishes slowly with height. At the uppermost levels of the troposphere (usually 200 or 100 mb) it may become a high aloft.

Cold-core barotropic high

POLAR AIR SOURCE REGION

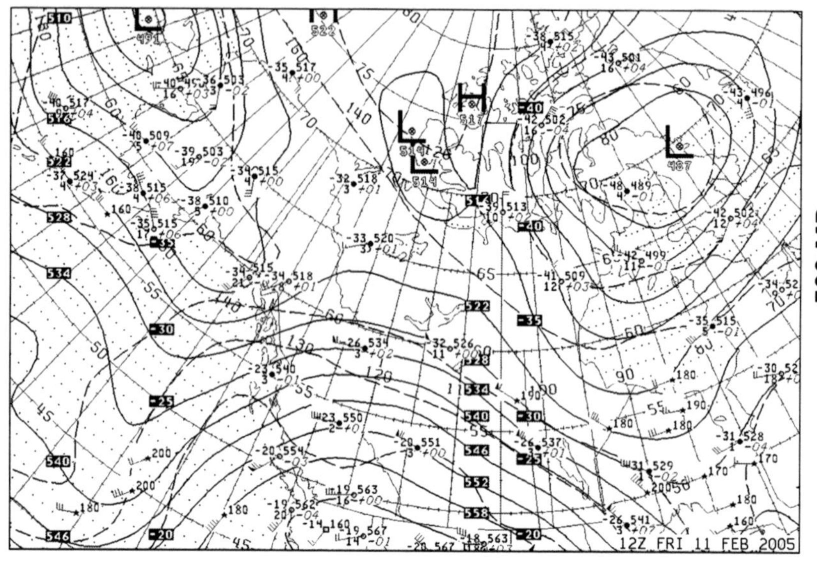

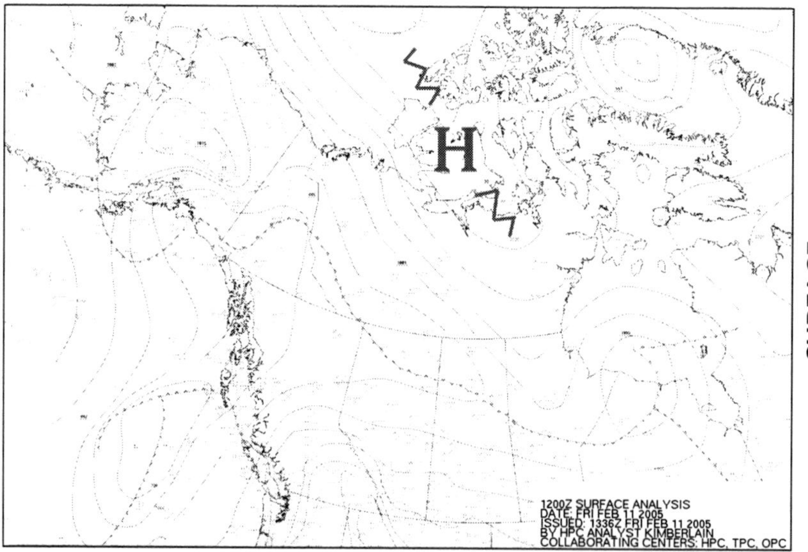

Figure C2-1g. Cold-core barotropic high (polar air source). A polar air mass in its source region appears as high pressure aloft stacking vertically into a weak low with height. The lowest heights are over the coldest part of the air mass. These systems tend to rapidly become baroclinic.

Cold-core barotropic high

PLATEAU HIGH

500 MB

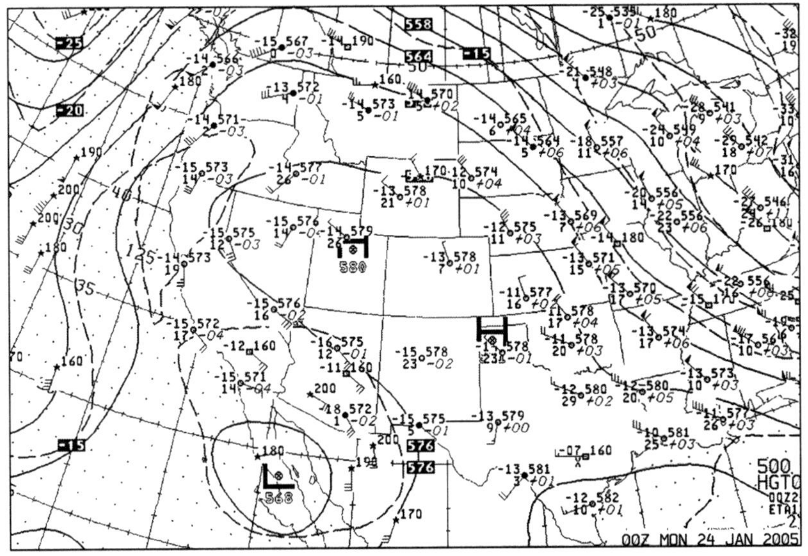

SURFACE

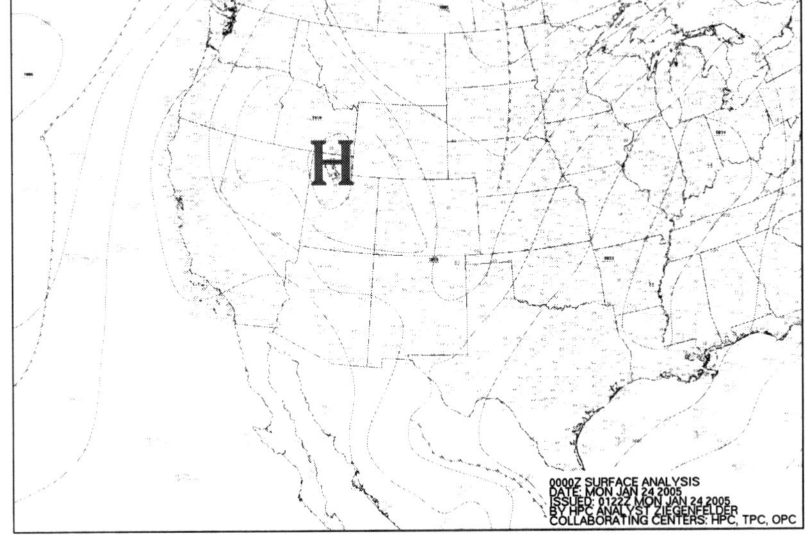

Figure C2-1h. Cold-core barotropic high (plateau high). The plateau high, which often occurs in the Great Basin region of the United States due to rapidly-cooling, stagnating air, is stacked vertically and weakens with height from a strong surface high into a weak high or weak low aloft.

Warm-core barotropic high

SUBTROPICAL HIGH

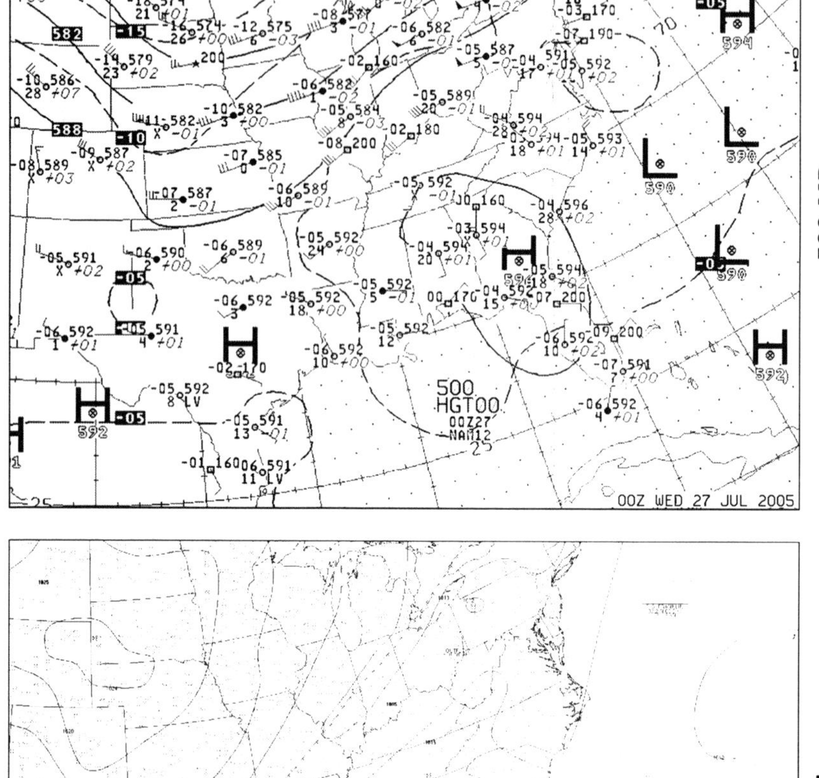

Figure C2-1i. Warm-core barotropic high (subtropical high). The subtropical high is formed from convergence aloft between the Hadley and Ferrel cells and the resulting subsidence. It stacks vertically (sometimes slightly toward the warm air) from a weak surface high to a moderate or strong high aloft.

Warm-core barotropic high

CUTOFF HIGH

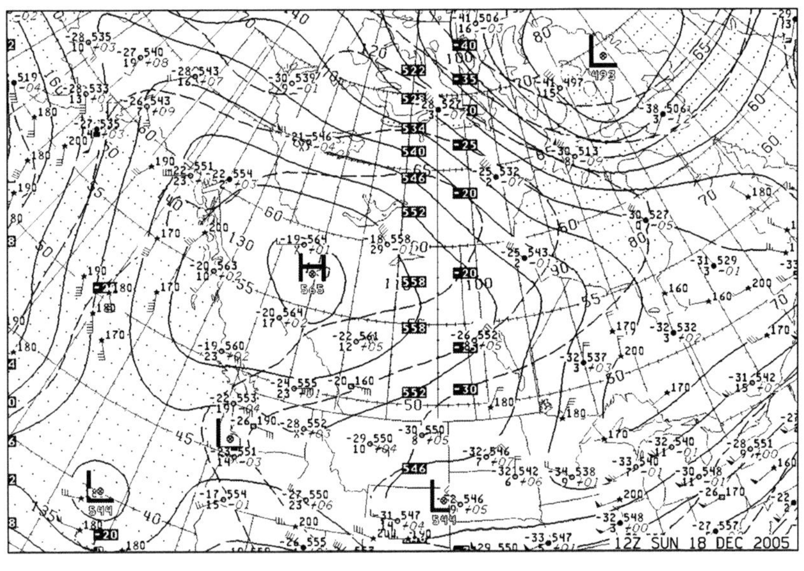

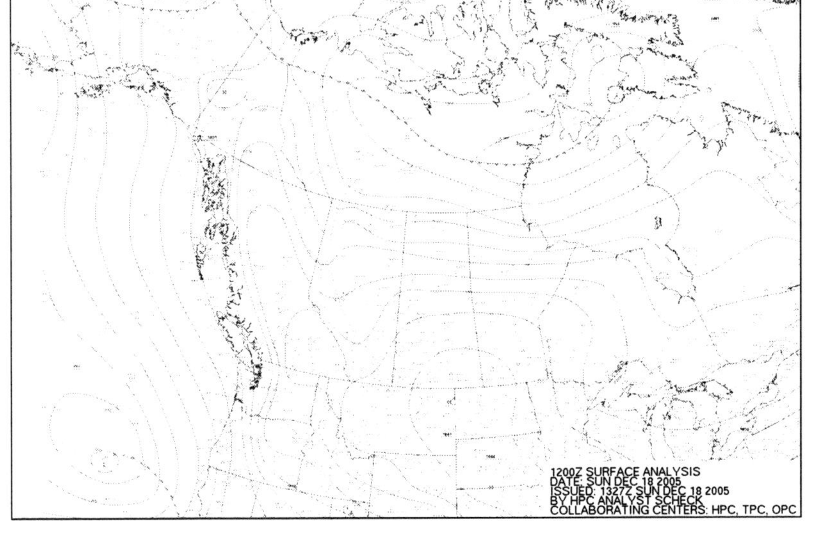

Figure C2-1j. Warm-core barotropic high (cutoff high). The cutoff high is formed when an upper-level ridge becomes detached poleward of the prevailing westerlies. It stacks vertically from a weak surface high into a strong upper-level high. It is most common during the spring, and often occurs in the Atlantic Ocean.

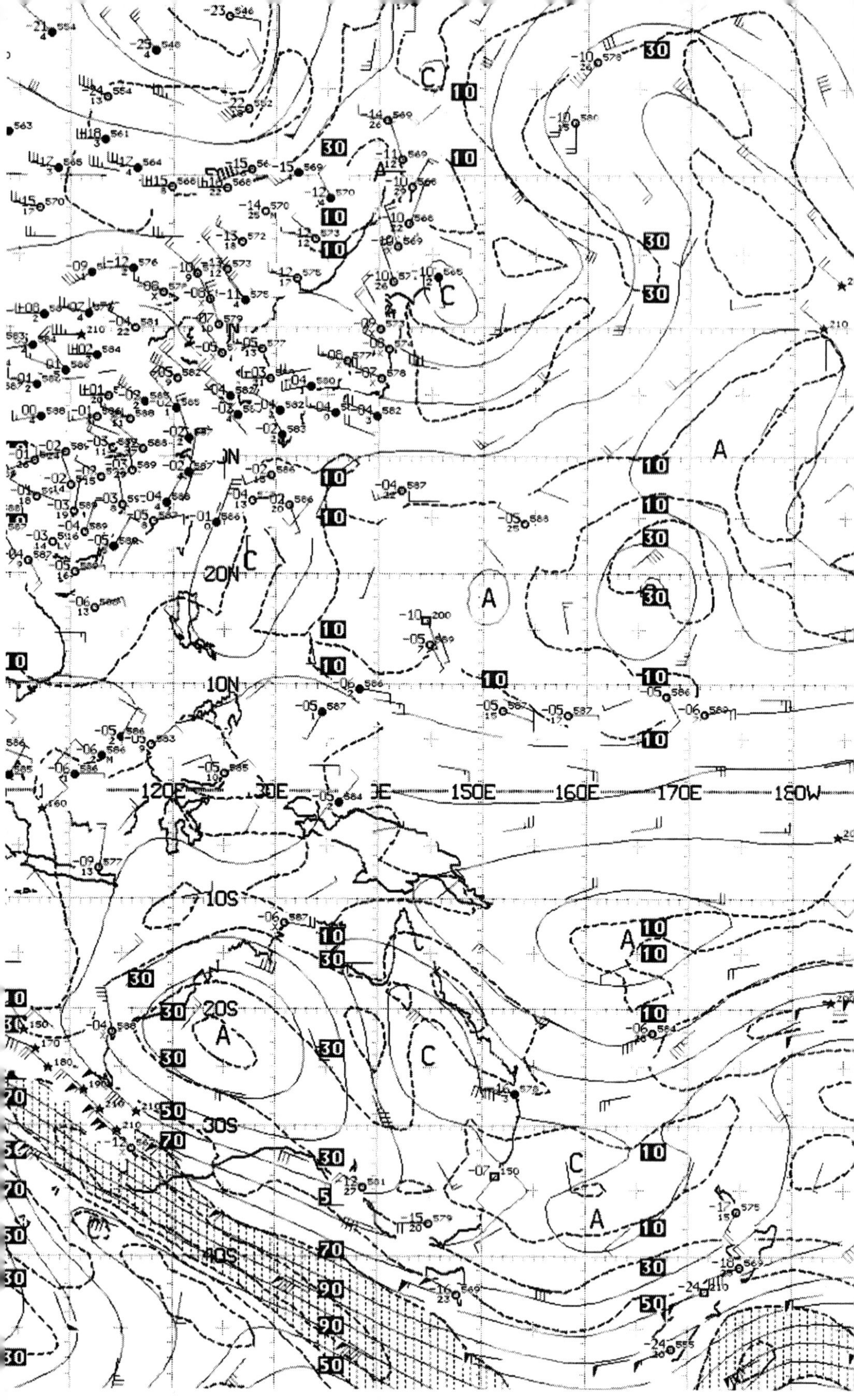

120E
150E
160E
170E
180W
10N
20N
10S
20S
30S
40S

C3. FORECASTING

Upper-level systems

1. Wave types

1.1. **Long wave**. A long wave is a continental-scale upper-level wave that has a wavelength of about 40 to 120° of longitude.

1.1.1. SEASONAL DISPLACEMENT. During the winter months, long wave troughs tends to cover cold continental areas with ridges offshore. During the summer months, long wave ridges tend to cover warm continental areas, with troughs offshore.

1.1.2. WAVE NUMBER. The wave number indicates the long wave count around the hemisphere. Low wave numbers are associated with stationary or retrogressing waves; high numbers are associated with progression. The influence of the long wave number: (Wolff, 1955)

3-4	Typically increases to 5 within 24 hours
5	Most common; blocking patterns often occur

1.1.3. CHANGE IN LONG WAVE PATTERN. Most height changes are related to short wave movement, but an unusually large geographic area of height changes may indicate a change in the long wave pattern. [ATC/1]

1.2. **Short wave**. Short waves are mesoscale or subsynoptic features embedded in the long wave patterns. A short wave trough may flatten a long wave ridge as it moves through the flow.

1.3. **Blocks**. Blocks are specific patterns which cause the long wave flow to become stationary for days or weeks at a time. Large barotropic systems such as cutoff lows may also serve as blocks.

1.3.1. OMEGA BLOCK is a ridge that is much broader on the poleward side than its equatorward side.

1.3.2. DIPOLE BLOCK or REX BLOCK is a closed high poleward of a closed low.

2. Upper troughs

An upper trough is any elongated area of low geopotential heights (pressure) in the upper troposphere.

2.1. STACK. The vertical axis of low pressure systems appear to *"stack" towards the coldest air with height.* This is an effect of the lower thicknesses (and implied lower heights) above the cold air mass.

2.2. TILT. The orientation of the trough axis has been shown by Macdonald (1976) and Glickman et al. (1977) to correlate with the incidence of thunderstorm activity.

- *Negative tilt.* When the axis of a trough extends northwest-southeast (southwest-northeast in the Southern Hemisphere), a negative-tilt trough is described. This correlates with a higher incidence of widespread thunderstorms and with larger precipitation amounts. Negative-tilt troughs are not as common as positive-tilt troughs.
- *Positive tilt.* When the axis of a trough extends northeast-southwest (southeast-northwest in the Southern Hemisphere), a positive-tilt trough is described. This correlates with a lower incidence of widespread thunderstorms. Positive-tilt troughs are more common than negative-tilt troughs.

2.3. DEEPENING. A mid-level trough may deepen when:

- moving toward a long wave trough
- a jet max is behind the trough rather than ahead of it
- low-level cold-air advection is occurring underneath

2.4. FILLING. A mid-level trough may fill when:

- moving away from a long wave trough
- a jet max is ahead of the trough rather than behind it
- low-level warm-air advection is occurring underneath
- the trough contains zonal flow in its rear. [FTW]

2.5. MODEL PERFORMANCE. Numerical models have historically moved upper troughs too fast and not deepened them enough.

3. Upper ridges

An upper ridge is any elongated area of high geopotential heights (pressure) in the upper troposphere.

3.1. STACK. The vertical axis of high pressure systems appear to *"stack" towards the warmest air with height.* This is an effect of the higher thicknesses (and implied higher heights) above the warm air mass.

3.2. INTENSIFICATION. A mid-level trough may intensify when:

- moving toward a long wave ridge
- a jet max is behind the ridge rather than ahead of it
- low-level warm-air advection is occurring underneath

3.3. WEAKENING. A mid-level ridge may weaken when:

- moving toward a long wave trough
- a jet max is ahead of the ridge rather than behind it
- low-level cold-air advection is occurring underneath
- the trough contains zonal flow in its rear. [FTW]

3.4. MODEL PERFORMANCE. Numerical models have historically moved upper ridges too fast and not built them up enough.

4. Upper lows

4.1. **Occlusions**.

4.1.1. MOVEMENT. Closed lows and/or cyclonic vorticity areas move parallel to the direction of maximum winds, but remain left (right in Southern Hemisphere) of the jet. [ATC/1]

4.1.2. HUDSON BAY VORTEX. A cold core low is frequently observed over eastern Canada (usually over Hudson Bay) during the winter months. If this low is displaced southward over the Great Lakes or New England, short waves moving through the circulation often produce intense storms with strong surface winds, cold advection, and extensive precipitation across the northeast U.S. (Weber, 1980)

4.2. **Cutoff lows**.

4.2.1. CLIMATOLOGY. Closed lows may remain in the southwest United States for several days or longer. An increase in precipitation over central and southern California occurs with this pattern and spreads eastward. (Weber, 1980)

4.2.2. HENRY'S RULE. Henry's rule states that a cold low aloft in the southwestern United States should not be forecasted to move until the katallobaric center [i.e. location of maximum pressure fall] within a trough aloft ["kicker"] to the northwest [usually over Pacific Northwest] comes within 1200 nm of the low. (Henry, 1949)

4.2.3. DEVELOPMENT. Southwest U.S. or southern Europe cutoff lows are indicated by (a) rapid intensification of the upstream ridge; (b) height falls moving south or southwestward on the back side of the associated trough. [ATC/1]

4.2.4. INTENSIFICATION. Closed lows will likely deepen if (a) divergent upper-level winds are occurring; (b) strongest winds or mean tropospheric cold advection are on the back side of a trough; (c) height falls are occurring to the rear of the trough axis; or (d) warming is noted in the stratosphere. [ATC/1]

4.2.5. HEIGHT FALLS. Closed lows will follow slightly to the left (right in Southern Hemisphere) of the track of the associated height fall centers. [ATC/1]

5. Upper highs

■ Upper highs are likely to develop over the eastern part of oceans in higher latitudes. A precursor to development is a sharply-curved ridge, with a jet max approaching the base of this ridge from the west; the jet "undershoots" the ridge. It may also occur if mean tropospheric warm advection is taking place poleward or poleward and to west on the back side of the ridge. [ATC/1]

■ Anticyclonic vorticity areas and/or upper-level highs move parallel to the direction of maximum winds, but remain right (left in Southern Hemisphere) of the jet.

■ Blocking upper highs may appear over Alaska during the winter months when the westerlies shift southward. (Weber, 1980)

6. Polar front jet

The polar front jet (PFJ) is applied to any jet formed where the baroclinic zone extends downward through the troposphere. The polar jet's mean position is equatorward during the winter and poleward during the summer. Wind speeds of a polar jet are normally higher in the winter than in the summer.

- LOCATION: Temperate regions directly above 500 mb front. May be found roughly near the 500 mb -17°C isotherm.
- HEIGHT: Between 300 and 250 mb; ~3000 ft below tropopause
- WIND VELOCITY: May exceed 200 kt in the winter months

7. Subtropical jet

The subtropical jet (STJ) occurs, as its name suggests, in subtropical latitudes. It is much more common during the winter months. *The baroclinic zone is restricted to the upper part of the troposphere*, with little direct connection between the jet and surface developments.

- LOCATION: Subtropical latitudes between 20 and 30 deg. May be found roughly near the 500 mb -11°C isotherm.
- HEIGHT: Usually 200 mb, but up to 150 mb in tropical latitudes
- WIND VELOCITY: Maximum winds may exceed 150 kt

8. Equatorial easterly jet

The equatorial easterly jet is a band of very strong *easterly* winds that occurs in the Asian and African tropics.

- LOCATION: Eastern hemisphere between equator and 20 deg N
- HEIGHT: 100 mb near 20 deg N to 200 mb near equator
- WIND VELOCITY: Maximum winds may exceed 120 kt

9. Polar-night jet

(Has little or no effect on surface weather)

A polar-latitude wintertime jet in the upper stratosphere. It is thought to occur when the stratosphere encounters perpetual darkness and a large cold pool results. The polar-night jet is above the strong temperature gradient on the edge of the pool.

- LOCATION: 60 to 70 degrees latitude
- HEIGHT: 1 mb (45-50 km)

- WIND VELOCITY: 364 kt has been recorded by sounding rockets

10. Nocturnal low-level jet

The nocturnal low-level jet (LLJ) is common in the southern and Great Plains regions of the United States. It has, however, been observed in the Carolinas, the New England coast, in western Europe, and in the Persian Gulf region. It is most pronounced during the late night hours, during the transition seasons, and during strong warm advection regimes. In the central U.S., the low-level jet often occurs when clear skies and light southeasterly surface winds are present (Gerhardt 1962). Decoupling of the air mass aloft due to the presence of an inversion (such as a radiational inversion or a cap) reduces frictional force, and may allow the winds above the surface to become supergeostrophic. When the forecast area is under the subsident right front quadrant of a jet streak, development of a nocturnal LLJ is favored.

- LOCATION: Temperate regions in warm air mass regimes.
- HEIGHT: Near 850 mb
- WIND VELOCITY: Strong low-level jets average about 50 kt

11. Regional

11.1. **Western U.S**. The appearance of ridges over the West Coast of the U.S. typifies a normal winter pattern. Occasionally a persistent cyclonic flow pattern over the West Coast extends far into the Pacific Ocean. In these cases, heavy rains and even floods may occur. The two main criteria are persistence of a long fetch of southwest flow and a good thermal gradient in the air mass. (Weber, 1980)

C4. FORECASTING

Vertical motion

Assessment of vertical motion is a critical part of weather forecasting. Clouds and rain are the direct result of ascent of a saturated air mass, while descent favors clearing weather.

1. Conceptual model

1.1. Upper level troughs and ridges.

1.1.1. AHEAD OF TROUGH. Cyclonic vorticity advection causes increasing cyclonicity. This causes midtropospheric divergence. This in turn decreases the mass of the column and lowers suface pressure. Upward motion occurs. This cools the lower troposphere adiabatically, decreasing thickness and leading to height falls. The trough propagates eastward as a result.

1.1.2. AHEAD OF RIDGE. Anticyclonic vorticity advection causes increasing anticyclonicity. This causes midtropospheric convergence. This in turn increases the mass of the column and increases suface pressure. Downward motion occurs. This warms the lower troposphere adiabatically, increasing thickness and leading to height rises. The ridge propagates eastward as a result.

1.2. Surface highs and lows.

1.2.1. AHEAD OF BAROCLINIC CYCLONE. Warm advection in the lower troposphere causes thickness to increase, thus upper-level heights rise. The height rises cause divergence aloft. Upward motion results, and the divergence also removes mass from the column and lowers surface pressure. The surface low propagates forward as a result.

1.2.2. BEHIND BAROCLINIC CYCLONE. Cold advection in the lower troposphere causes thickness to decrease, thus upper-level heights fall. The height falls cause convergence aloft. Downward motion results, and the convergence also adds mass to the column and increases surface pressure. The surface low propagates away as a result.

1.3. **Self-development**. The upper-level and surface development cycles reinforce each other, causing a chain reaction known as *self-development*. Self-development is hindered only by friction and diabatic processes. In extreme cases, self-development over ideal regions such as the Gulf Stream waters during winter may cause the creation of rapidly-deepening systems known as *bombs*.

1.4. **Jet streak quadrants**. For purposes of brevity all of these conceptual models are valid for the Northern Hemisphere; for use in the Southern Hemisphere they should be inverted. Air in geostrophic balance entering the entrance region (rear, typically west side) of a jet streak encounters a region of tighter pressure gradients. Pressure gradient force (PGF), which acts to the left of a parcel's motion, becomes dominant, so parcels become ageostrophic and accelerate to the left. As their velocity increases, Coriolis force increases and re-establishes the balance. The parcels move through the jet streak in balance. As they exit the jet streak in the exit region (front, typically east side), pressure gradients decrease; PGF becomes weaker and Coriolis force, which acts to a parcel's right, becomes dominant. The parcels become ageostrophic, move towards the right, and decelerate. As their velocities decrease, Coriolis force diminishes and the parcels are back in balance. These ageostrophic unbalanced areas produce convergence and divergent areas, and as a result since the jet is in the upper troposphere, vertical motion must exist to compensate.

1.4.1. STRAIGHT FLOW. The above mentioned ageostrophic flow causes convergence in the *right rear* and *left front* quadrants, where compensating ascent occurs to transport mass from the lower to the upper troposphere. There is also divergence in the *left rear* and *right front* quadrants, where compensating subsidence occurs to transport mass from the upper to the lower troposphere.

1.4.2. CYCLONICALLY CURVED FLOW. The relation of jet streak quadrants for straight flow applies. Cyclonically-curved flow, in which flow curves to the left, causes *enhanced vertical motion on the left side* and *weak vertical motion on the right side*. Thus the rising left front quadrant and subsident left rear quadrant (inside the curve) are of primary importance.

1.4.3. ANTICYCLONICALLY CURVED FLOW. The relation of jet streak quadrants for straight flow applies. Anticyclonically-curved flow, in which flow curves to the right, causes *enhanced vertical motion on the right side* and *weak vertical motion on the*

Figure C4-1. Jet quadrant conceptual model, with variations for straight-line and curved flow. These examples are applicable to the Northern Hemisphere.

STRAIGHT-LINE FLOW

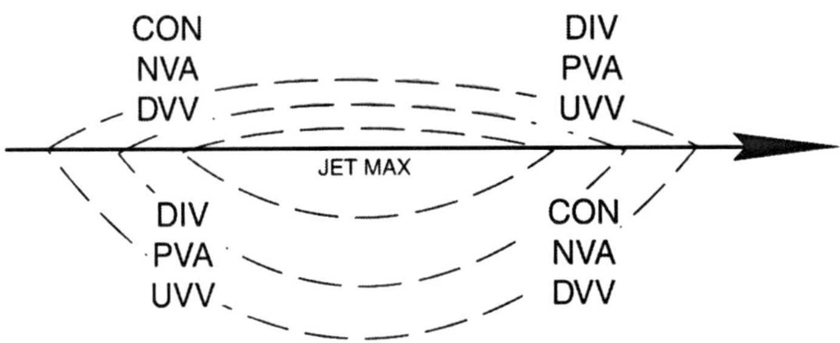

CYCLONICALLY-CURVED FLOW

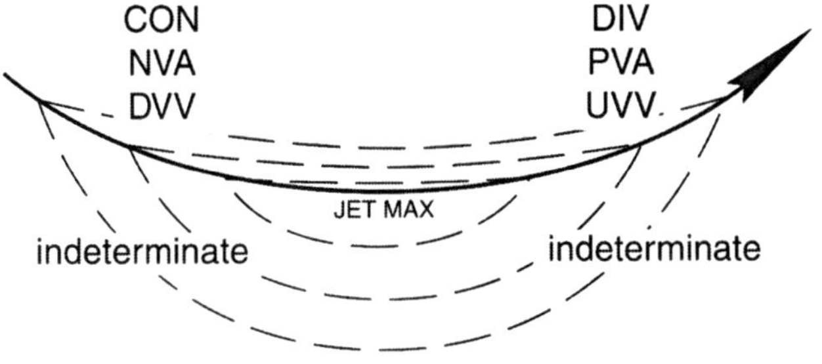

ANTICYCLONICALLY-CURVED FLOW

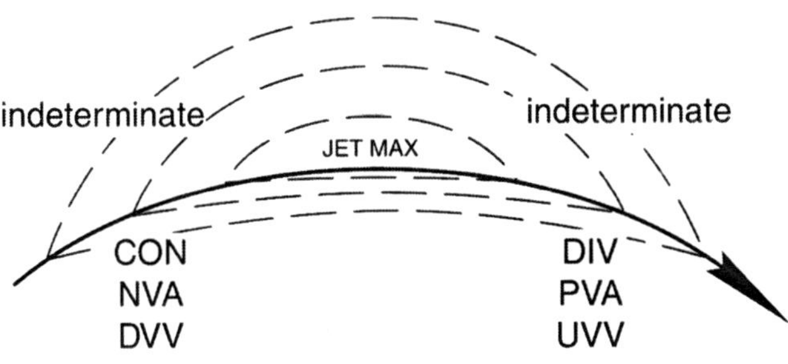

left side. Thus the rising left front quadrant and subsident left rear quadrant (inside the curve) are of primary importance.

2. Vorticity relationship

The technique of using the 500 mb vorticity chart to assess synoptic-scale vertical motion was introduced in 1952, and this method came into widespread use among U.S. forecasters during the 1960s. It is by vorticity advection that a steady-state quasigeostrophic system moves. Interestingly many of the convergent and divergent areas indicated by the jet streak quadrants are reflected by vorticity fields.

2.1. **Vorticity patterns.** When areas of high vorticity are advected to a particular location, positive vorticity advection is said to be taking place. Since height contours approximate the wind field, the crossing of vorticity contours against height contours is a de facto indicator that this is occurring. If numerous vorticity contours cross the height contours, stronger advection is indicated.

2.1.1. RELATION TO TROUGHS. Troughs contain *cyclonic vorticity* (*positive vorticity* in the Northern Hemisphere). Where wind is flowing out of the trough, *cyclonic vorticity advection* (CVA) (*positive vorticity advection* in the Northern Hemisphere) is occurring. This is usually on the east side of the trough, but may be on the west side of the trough if the wind flow is east-to-west or in rare instances where the trough is moving faster than the environmental winds.

■ *Advection lobe*. If a short wave trough axis is perpendicular to the height fields (Figure C4-2a), vorticity contours will show sharp advection patterns. It is indicative of a short wave

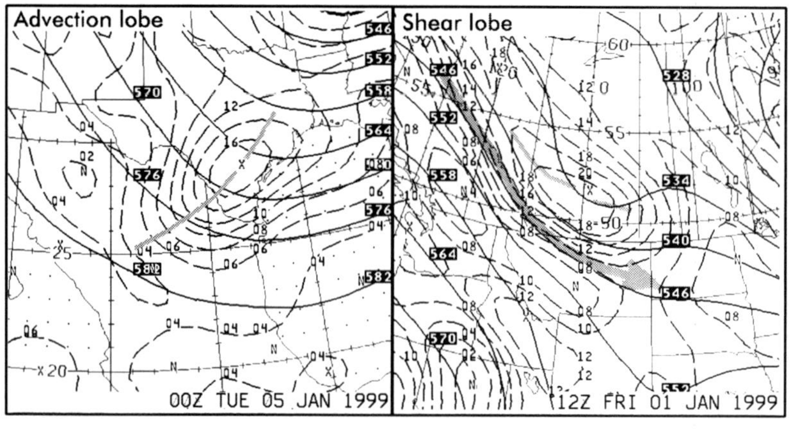

Figure C4-2. Advection and shear lobes.

embedded in the flow. The associated jet is sometimes called an *advection jet.*

■ *Shear lobe.* If a short wave trough axis is parallel to the height fields (Figure C4-2b), vorticity contours will show poor advection patterns. This is indicative of a jet streak embedded in the flow. The CVA and AVA patterns will tend to coincide with that of the jet streak conceptual model. The area is sometimes called *channeled*, with the associated jet called a *channel jet.*

2.1.2. RELATION TO RIDGES. Ridges contain *anticyclonic vorticity* (*negative vorticity* in the Northern Hemisphere).Where wind is flowing out of the ridge, *anticyclonic vorticity advection* (AVA) (*negative vorticity advection* in the Northern Hemisphere) is occurring. This is usually on the east side of the ridge, but may be on the west side of the ridge if the wind flow is east-to-west or in rare instances where the ridge is moving faster than the environmental winds.

2.1.3. BOX METHOD. A “box method” is often used mentally or graphically to assess vorticity advection. Boxes are drawn, shaped by the intersection of vorticity contours against height contours (Figure C4-3). Where the *boxes are small and tightly packed, strong advection* is indicated. Where the *boxes are large and spread out, weak advection* is indicated.

2.2. **Vertical motion.** The ascent or descent of air is tied to what type of vorticity advection is occurring. Typically the 500 mb chart is used.

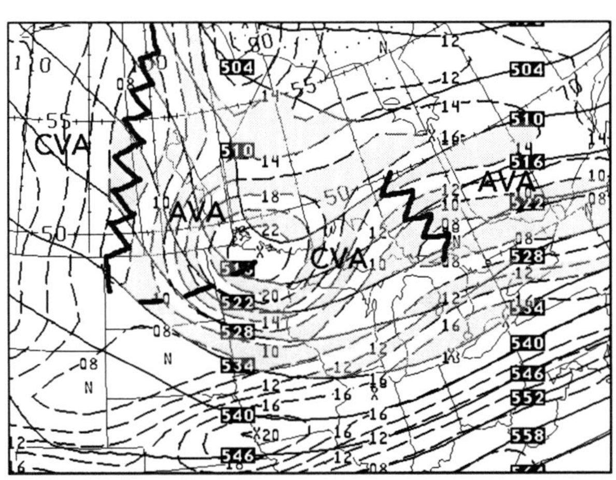

Figure C4-3. Box method of determining vorticity advection, with short wave trough (dashed line) and ridges (zigzag lines) superimposed. The smallest, densest areas of boxes correspond with the strongest vorticity advection. In this case, the best CVA is well ahead of the short wave trough.

2.2.1. RISING MOTION. Divergence will occur where cyclonic vorticity advection (CVA) (positive vorticity advection in the Northern Hemisphere) is taking place.

2.2.2. SINKING MOTION. Conversely, convergence will occur where anticyclonic vorticity advection (AVA) (negative vorticity advection in the Northern Hemisphere) is taking place.

2.2.3. ASSUMPTIONS. A diagnosis of vertical motion based on vorticity advection assumes the following:

- *An increase with height of vorticity values* due to the normal increase in wind speed with height. If winds are decreasing through the level of vorticity analysis, the inverse type of vertical motion will occur.

- *Vertical motion is not offset by thermal advection* (see Isentropic Diagnosis section below). Warm air advection below is associated with lift, and cold air advection is associated with subsidence. It is very common for thermal advection to negate the vertical motion implied by a vorticity pattern.

2.3. **Trenberth's method** shows that advection of vorticity in the middle of a layer by the thermal wind throughout its depth closely represents the quasigeostrophic forcing for vertical motion (Trenberth 1978). Superimposing traditional 500 mb vorticity contours against 1000-500 mb thickness contours and performing "box counting" approximates this method, the primary shortfall being that 500 mb vorticity is at the top of the layer being sampled.

2.4. **Limitations**. The assumption of vertical motion depends on the thermal advection taking place in the layers underneath. Warm advection implies rising motion, while cold advection implies sinking motion.

3. Omega

A diagnostic equation for vertical motion was presented by Holton (1972). It uses the change in vorticity advection *with height* and the Laplacian of thermal advection. Since both of these quantities frequently cancel each other out and are difficult to compute accurately, the omega equation is, in practice, an estimate at best.

- **Positive omega** (mb/s^{-1}): Subsidence
- **Negative omega** (mb/s^{-1}): Ascent

Special note: t has been noted that some sources use positive omega to show ascent. When using a source of omega products for the first time it is prudent to crosscheck the fields before making interpretations (e.g. select a major short wave trough and check the mathematical sign of the omega field ahead of and behind it).

4. Q-vectors

The Q-vector approach was, by the 1980s, considered a superior method for diagnosing vertical motion. It was not well-adopted for a number of reasons. First, it can *only be calculated accurately from observed data*, not from model output, and forecasting tools during the past few decades have gravitated away from raw observational datasets and towards centrally-produced fields. Also there is often ambiguity about which level Q-vectors should be viewed at.

4.1. **Components of Q-vectors**. Q-vectors are made up of two components: **Qs** (the rotational component) which is aligned parallel with thickness/isotherms, and **Qn** (the frontogenetic component) which is aligned perpendicular to thickness/isotherms.

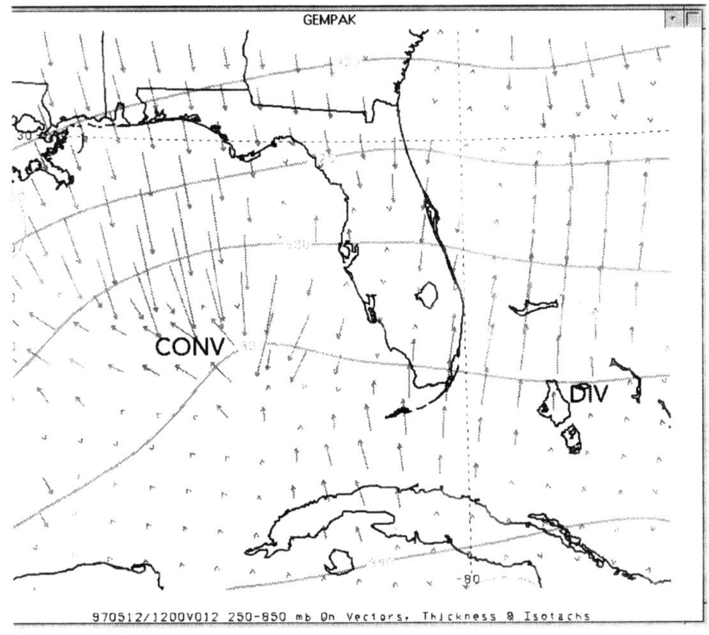

Figure C4-4. Q-vector graphic showing upward motion and frontogenesis in the region west of Florida. East of Florida there is frontolysis and subsidence.

4.2. **Interpretation of vertical motion**. Vertical motion is indicated by the divergence or convergence of Q-vectors (Figure C4-4). The strongest vertical motion is at the center of this field.

4.1.1. DIVERGENCE. If there is divergence of the Q-vectors, subsidence is implied. *(Positive divQ)*

4.1.2. CONVERGENCE. If there is convergence of the Q-vectors, ascent is implied. *(Negative divQ)*

4.3. **Frontal diagnostics**. If Q-vectors cross thickness lines (i.e. if the Qn component is strong), frontogenesis or frontolysis is indicated. The stronger the Qn component, the greater the change in the thermal gradient with time. The findings are representative of the Q-vector layer being studied; thus the 1000-850 mb layer or 850-700 mb layer is best for analyzing frontal zones.

4.2.1. FRONTOGENESIS. Frontal intensification is indicated when Q-vectors point towards warm air.

4.2.2. FRONTOLYSIS. Frontal dissipation is indicated when Q-vectors point towards cold air.

5. Isentropic analysis

Isentropic forecasting assumes that air flowing “horizontally” has a tendency to cling within its own isentropic surface. Isentropic surfaces are deformed *upwards by cold air* and *downwards by warm air*.

6.1. **Diagnosis of vertical motion**. Given the deformation of isentropic surfaces upward over cold air, the following relationship can be inferred when viewing motion on an isentropic surface:

6.1.1. ASCENT. When isentropic flow is towards low pressure (high heights), lift is indicated.

6.1.2. SUBSIDENCE. When isentropic flow is towards high pressure (low heights), subsidence is indicated.

6.2. **Diabatic processes**. It must be noted that air is **not** always constrained to move along isentropic surfaces, especially in diabatic conditions such as surface heating, precipitation, and so forth.

C5. FORECASTING

Fronts

1. General

Fronts are located on the *warm side* of all transition zones (both the transition zone as indicated by an inversion on the sounding, and the transition zone from cold to warm air as seen on surface and upper-level charts). All fronts are located in a pressure trough, and winds veer with frontal passage. Fronts aloft are found towards the cold air mas with increasing height.

1.1. **Fronts tend to intensify** (frontogenesis) if: [ATC/ER]
- (a) air mass contrast is increasing;
- (b) the front approaches a deep pressure trough; or
- (c) the front changes from inactive (katafront) to active (anafront).
- (d) the thickness gradient is increasing

1.2. **Fronts tend to dissipate** (frontolysis) if: [ATC/ER]
- (a) air mass contrast is decreasing;
- (b) the front is moving into higher pressure; or
- (c) the front changes from active (anafront) to inactive (katafront).
- (d) the thickness gradient is decreasing

1.3. **Frontal precipitation**.
- Frontal precipitation will be more intense when the angle of isobars and wind shifts across the front indicate strong convergence. [ATC/ER]
- In subtropical latitudes, cold fronts produce more precipitation; whereas in polar latitudes, more precipitation accompanies warm fronts and occlusions. [ATC/ER]

2. Cold front

A cold front is where warm air is being replaced with cold air. The mean winds normal to the front usually blow from cold air to warm air.

2.1. **Speed**. Cold fronts will move at 85% of the second standard level winds (about 1000-2000 ft AGL) in the cold air. [ATC/ER]

2.2. **Cold front type**. Cold fronts are sometimes classified as "anafront" or "katafront", depending on the mean wind tropospheric wind flow perpendicular to the front. The difference between anafronts and katafronts is dramatic and their impact can be significant. One should check the backing of the winds [vertically behind the front], the strength of the frontal inversion, and the presence of dry/moist layers to help classify a cold front as an anafront or katafront. (Moore and Smith 1989).

2.2.1. ANAFRONT or ACTIVE FRONT. This type of front is marked by a sudden, large temperature and slight humidity drop with frontal passage, slow clearing with frontal passage, moderate to heavy rain with frontal passage giving way to steady postfrontal rain, and a sharp wind shift followed by decreasing winds behind the cold front (Sansom 1951). The extensive postfrontal cloudiness and precipitation can enhance rain/snow totals, radically alter maximum and minimum temperatures, and cause prolonged icing conditions for aircraft (Moore and Smith 1989).

2.2.2. KATAFRONT or INACTIVE COLD FRONT. This type of front brings a slight temperature drop and sharp relative humidity drop with frontal passage; rapid cloud clearing after frontal passage; very light rain in a narrow band along the front with possible convection ahead of the front; and a gradual change in winds with slight speed changes after the front passes (Sansom 1951). Katafronts tend to be associated with prefrontal precipitation and clear, dry conditions behind the front; more typical of the "textbook" cold front (Moore and Smith 1989).

3. Warm front

A warm front is where cold air is being replaced with warm air. The wind normal to the frontal boundary flows from warm to cold air. If the cold air mass is retreating as rapidly as the warm air is advancing, there is little lift along the frontal surface; however if the cold air is retreating slowly, then lift and possibly widespread precipitation will develop.

3.1. **Movement**. Warm fronts tend to move at 70% of the second standard level winds (about 1000-2000 ft AGL) in the cold air ahead of the front. [ATC/ER]

3.2. **Precipitation**. The zone of precipitation ahead of a warm front will be narrow if the axis of the surface high is close to the warm front. [ATC/ER]

4. Quasistationary front

A quasistationary (stationary) front exhibits little or no movement. Weather along the front is similar to that of a warm front.

5. Occluded front

An occluded front results when two fronts (involving three air masses) merge. This often happens around cyclones where the cold front catches up to the warm front. The air mass with the lesser density is driven aloft and a single front results.

5.1. **Occlusion type**. Once an occlusion has developed, its type depends on the characteristics of the air ahead of and behind it:

5.1.1. COLD OCCLUSION. If the air behind the occluded front is *colder* than the air ahead of it, it will act as a cold occlusion. This is the most common type of occlusion.

5.1.2. WARM OCCLUSION. If the air behind the occluded front is *warmer* than the air ahead of it, it will act as a warm occlusion. Such occlusions are rare but they do occur, often along the coast of Washington and British Columbia where Pacific systems advance towards much colder air masses.

6. Dryline

A dryline delineates a transition zone marked primarily by a strong dewpoint contrast. Traditionally the dryline separates maritime tropical (mT) air from continental tropical (cT) air.

6.1. **Movement**. The dryline *advances* towards the moist air during the day; the mechanism is the progressive mixing of moist air into the free atmosphere. The dryline usually *retreats* towards the dry air at night. The mechanism is advection of moist air as mixing is suppressed. The diurnal-nocturnal movement cycle is sometimes referred to as *sloshing*. Very deep moist air or limited heating will result in poor advancement of the dryline during the day, while shallow moist air or strong heating will result in fast advancement. Nighttime movement is generally a function of the pressure gradient across the dryline (e.g. strong pressure falls in the dry air mass will result in a fast retreat).

6.2. **Role in convection**. Severe storms may form along the dryline. Typically this is because the western edge of the dryline coincides with the most favorable environment for severe weather. Convergence of the wind field along the dryline may enhance prospects for convective initiation.

7. Regional

The following techniques were developed for use in specific forecast locations.

7.1. **Texas**.

- When a low is in Colorado, a Pacific front will not pass Fort Worth until the pressure is higher at Salt Lake City than in Brownsville, Texas. [FTW]
- A Pacific front will move to the upper Texas coast in about 24 hours once the pressure in Salt Lake City exceeds that of New Orleans. [FTW, Ray T. Telfer]

C6. FORECASTING

Wind

1. Wind phenomena.

1.1. **Chinook wind**. Three types of chinooks have been categorized in Montana (Price 1971).

1.2.3. FRONTAL CHINOOK. A common type of chinook associated with an eastward-moving cold front crossing the Rockies. The surface gradient across the Rockies strengthens, usually *ahead* of the cold front

1.2.1. BASIN HIGH CHINOOK. Occurs during the cold season when a stagnant Great Basin high (southwest of Montana) underneath an upper level ridge is affected by a strengthening cross-mountain surface gradient. Surface temperatures may rise into the 50s and 60s.

1.2.2. KLONDIKE CHINOOK. A rare type of chinook that may follow an arctic outbreak as warm advection from the west erodes and displaces the arctic air. Temperatures rise only into the 20s, causing a wind chill during the episode, and blowing snow may occur.

2. Regional Techniques

It is important to note that different physics should not be expected in various areas; rather it is the unique geography and topography, and the frequency of certain weather patterns that lead to these techniques and rules of thumb.

1.1. **Southern California**.

1.1.1. SANTA ANA WINDS. These winds blow from the northeast, associated with strong pressure gradients between Great Basin highs (e.g. Nevada) and lower pressure off the southern California coast.

1.2. **Montana**. The following equation was devised by Oard (1993) to forecast peak wind occurrence at Great Falls, Montana. It can be adjusted for other locations.

$$V_p = 7.31 + 0.036a + 0.051b + 0.021c + 0.021d$$

where V_p is the maximum wind speed in meters per second; a is the 700 mb height difference, in meters, between Boise, Idaho and Edmonton, Alberta; b is the sea-level pressure change difference, in tenths of millibars, between Idaho Falls, Idaho and Helena, Mont.; c is the 700 mb height difference, in meters, between Medford, Ore. and Prince George, B.C.; d is the difference between the 12-hour 700 mb height change, in meters, at Lander, Wyo. and Great Falls, Mont.

C7. FORECASTING

Clouds

Clouds require a saturated air mass with a parcel relative humidity of 100% (the air temperature equals the dew point temperature). If the relative humidity drops below 100%, cloud droplets will tend to evaporate.

1. Cloud amount

1.1. **Instant humidity method**. Where no other information exists, clouds may be suspected at any layer where the relative humidity is 70% or greater or the dewpoint depression is 5 Celsius degrees or less.

1.2. **Expanded humidity method**. This may be used when nothing is available except layer humidity. It was originally developed to assess cloud cover from LFM and NGM tabular output. [MT]

RH	Cloud cover
<65%	0 oktas
70%	1-2 oktas
75%	3-4 oktas
80%	4-5 oktas
85%	6-7 oktas
>90%	8 oktas

1.3. **Dewpoint depression method**. If only dewpoint depression is available, as a last resort the following can be used to obtain cloud cover [MT]. This works only for low cloud types.

DD (deg C)	Cloud cover
0 to 2	OVC
2 to 3	BKN to SCT
3 to 4	SCT
4 to 5	SCT to FEW
more than 5	CLR

1.4. **Extrapolation**. Simply advect clouds downstream. This may be done using satellite observations, or even surface observations.

1.5. **Radar**. The WSR-88D and similar high-sensitivity radar images can be used to track clouds. It is important to use base reflectivity images from different levels to assess the direction, distance, and height of clouds. Most reflectivities on the WSR-88D are in the range of -12 to +15 dBZ.

2. Height

2.1. **Humidity technique for cloud height.** If no other information is available, cloud layer heights may be determined as follows:

2.1.1. CLOUD BASE. The cloud base is at the height above which the the relative humidity is more than 70% or the dewpoint depression is less than 5 °C (9 °F).

2.1.2. CLOUD TOP. The cloud top is at the height above which the relative humidity is less than 70% or the dewpoint depression is more than 5 °C (9 °F).

2.2. **Convective height calculations**. In a convective weather situation, in which it is daytime and the low cloud type consists entirely of cumuliform clouds, the bases of clouds can be forecast as follows:

2.2.1. SOUNDING METHOD. On the sounding, find where the dry adiabat, originating from the surface temperature, and the mixing ratio line, originating from the surface dewpoint, intersect. Adjust the temperature and dewpoint for the current or forecast values before using them. This is the lifted condensation level. It corresponds to the base of cumulus clouds.

2.1.2. DEWPOINT DEPRESSION METHOD. Obtain the dewpoint depression in degrees Fahrenheit, and multiply by 220. This is the rough height of the cloud base in feet. To use Celsius, multiply by 400 instead.

3. Type

3.1. **Convective cloud forecasting**.

3.1.1. WEAK CONVECTION. Cumulus clouds will grow and either evaporate or form small layers of stratocumulus or altocumulus. Moderate convection into a very strong inversion (such as a trade wind inversion or Great Plains cap) will produce extensive sheets of stratocumulus or altocumulus, especially if those layers are humid.

3.1.2. STRONG CONVECTION. Cumulus clouds will grow into cumulonimbus clouds. The anvil from this cloud will produce extensive cirrus which may persist for days after the storm has dissipated. Mid-level inversions will produce altocumulus clouds. If the mid- and upper-levels of the atmosphere are humid, these cloud forms will persist for many hours or days.

3.2. **Synoptic-scale upper-level lift**. Widespread lifting will produce mostly stratiform cloud types. Around the periphery of the system, cirrostratus and altostratus clouds are favored. Where greater lift is present, dense altostratus and nimbostratus clouds occur. Rainfall will humidify the layer near the ground and produce low-hanging stratus ("scud", "fractus") clouds, which may coalesce and obscure the nimbostratus. Upper-level lift combined with instability tends to produce stratocumulus types.

4. Areal coverage

4.1. **Gayikian advective cirrus method**. Using a current and forecast chart for the upper troposphere and examining changes to the wavelength and amplitude of the upper-level jet, the Gayikian method can be used to forecast changes in the area and density of cirrus clouds. *See Figure C7-1 for details.* [MT]

5. Regional techniques

5.1. **California**. A technique developed by Felsch & Whitlach (1993) identified several prerequisites for stratus surges on the central California coast: (1) a continuous stratus-free area out to 60 nm offshore between Point Conception and Shelter Cove at 0000 UTC; (2) *no* cold air cumulus development within 300 nm of the clear zone; (3) a clear zone that does not extend south of the Mexican border; (4) an Oakland 850 mb 0000 UTC temperature of greater than 9.8 deg C; and (5) a pressure difference between San Francisco (SFO) and Santa Maria (SMX) of –0.2 mb or less.

Figure C7-1. Gayikian's upper-level forecast technique to predict cirrus coverage.

WAVELENGTH	AMPLITUDE	EFFECTS
No change	Increasing	Cirrus spreads poleward, diminishes slightly equatorward; becomes more dense.
No change	No change	Cirrus will exist in the same location relative to the jet.
No change	Decreasing	Entire cirrus area contracts and becomes less dense.
Increasing	Increasing	Cirrus spreads poleward and to east; little change in density.
Increasing	No change	Cirrus elongates east-west; contracts north-south; becomes less dense
Increasing	Decreasing	Cirrus dissipates or coverage decreases; becomes less dense.
Decreasing	Increasing	Cirrus will contract but spread poleward, with slight density decrease.
Decreasing	No change	Cirrus will decrease in the eastern portion with no change in density.
Decreasing	Decreasing	Cirrus area will contract.

3.2. **Texas**.

- Stratus will clear out of north central Texas within 24 hours after the wind at Guadalupe Pass becomes westerly at 35 to 40 kt. If a low is forming/moving south of this area, clear 8 to 12 hours slower. [FTW]
- In southerly flow, when the air temperature is lower than the nighttime dewpoints 180 miles upstream, then stratus will form and persist most of the day. Temperatures will warm only about 12 degrees. [FTW]
- Cold air stratus will persist in north Texas as long as the temperature in central Oklahoma is as cold or colder than the dewpoint at Fort Worth. [FTW]

C8. FORECASTING

Fog

Fog is the name given to any cloud that exists at the Earth's surface. It rarely extends to a depth of more than a few hundred feet.

Fog intensity is a function of droplet size and the amount of condensed water vapor. However unusually thick fog can occur when combustion particles are present in the atmosphere.

1. Radiation fog

This is the "classic" form of fog that appears overnight with light winds and clear skies. At night as the Earth radiates its heat into a clear sky, the layer of air near the ground is cooled by conduction. If this layer is sufficiently moist, especially from ground moisture due to recent rains, fog will form. Snow has an inhibiting effect on fog formation since water vapor sublimates on snow.

1.1. **Fog criteria**. Indicators for fog formation are:

- Clear skies (or rapid clearing at night)
- Moist ground (the more moisture, the higher the probability)
- Light but not calm winds
- Suppressed daytime heating
- A late-afternoon dewpoint depression of 20 °F or less.
- Pollution, which may allow fog to develop before the relative humidity reaches 100%.

1.2. **Wind speed**. Wind speeds have a significant influence on the type and extent of fog formation:

1.2.1. NEAR-CALM WINDS (0-2 kt) will result in heavy dew or frost rather than fog since the chilled layer adheres within a few inches of the earth's surface.

1.2.2. VERY LIGHT WINDS (3-5 kt) will result in fog since the chilled layer gets mixed through a deeper layer.

1.2.3. LIGHT WINDS (6-10 kt) is associated with deep fog if moisture is rich and cooling is strong.

1.2.4. MODERATE WINDS (10 kt or more) results in turbulent mixing and is associated with stratus layers close to the ground rather than fog.

1.3. **Rules of thumb**. Listed here are some rules of thumb developed for fog forecasting:

- Frost tends to inhibit the formation of radiation fog; however as the frost melts, usually within an hour or two after sunrise, fog can and often does form very rapidly. [EUR]
- Studies by the U. S. Air Force in Europe have shown that 850 mb winds in excess of 15 kt is an excellent predictor for the *non-occurrence* of fog. [EUR]
- In Europe, where dry air seldom follows frontal passage, radiation fog often forms the first night after a front passes, and if ridging follows, the fog can persist until the ridge is replaced by another system. [EUR]
- A fog regime often ends with a transition from anticylonically- to cyclonically-curved flow. [EUR]

1.4. **Fog formation prediction**. By the 21st century, mesoscale numerical models began to demonstrate superior ability to predict fog over manual techniques (Golding 2002). However, manual techniques in the hands of an experienced local forecaster can provide excellent results. Interestingly a large share of the techniques originate from Europe, where fog is exceedingly common and has a great impact on aviation and commerce.

1.4.1. CROSSOVER TEMPERATURE. The lowest dewpoint temperature observed at the time of maximum afternoon heating is known as the crossover temperature. When the nighttime surface temperature drops to this value, fog may form. The technique yields an indirect inference about the vertical structure of low-level moisture. The crossover temperature method does not work when significant temperature or moisture advection is expected.

1.4.2. BRITISH QUICK FOG POINT (BQFP). The BQFP is identical to crossover temperature, however it forces the subtraction of two Celsius degrees from the dew point temperature observed at maximum heating.

1.4.3. FOG THREAT INDEX. This is computed by subtracting the fog point from the 850 mb wet-bulb potential temperature (the 850 mb wet bulb temperature dropped moist-adiabatically to the station level). Values above 3 indicate a low likelihood, and those below zero indicate a high likelihood.

1.4.4. FOG STABILITY INDEX (FSI). This was developed by the U.S. Air Force (Strauss 1979) for German forecasting in the late 1970s. The original formula is as follows:

$$F = 4T_{SFC} - 2\,(T_{850} + T_{d\,SFC}) + W_{850} \qquad \text{(Strauss 1979)}$$

where F is the Fog Stability Index (FSI), W is the wind speed in kt, T is the temperature in deg C, and T_d is the surface dewpoint in deg C. The most recent sounding from nightfall should be used. A FSI of greater than 55 indicates a low probability, and less than 31 indicates a high probability.

Other variations in use that will produce a different range of results include:

$$\mathrm{F} = (T_{SFC} - T_{850}) + (T_{SFC} - T_{d\,SFC}) + W_{850} \qquad \text{(source unknown)}$$

or from Wantuch 2002:

$$\mathrm{F_H} = 2\,|\,T_{SFC} - T_{850}\,| + 2\,(T_{SFC} - T_{d\,SFC}) + 2\,W_{850}$$

$$\mathrm{V} = -1.33 + 0.45 \times \mathrm{F_H}$$

where V is visibility in km.

1.4.5. SAUNDERS FOG POINT METHOD. Using the observed dewpoint D at the lifted condensation level during maximum heating, follow the mixing ratio line to the surface. The temperature shown is the fog point temperature. If the dewpoint profile dries rapidly with height, extrapolate that part of the dewpoint line below this level to find D.

1.4.6. CRADDOCK & PRICHARD METHOD. The formula for the Craddock & Prichard method is:

$$T_f = 0.044T + 0.844T_d - 0.55 + A$$

where T_f is the fog point, T is the temperature at midday, T_d is the dewpoint at midday, and A is an adjustment. As these were designed to use the 1200 UTC (1200 LST) sounding in Great Britain, possibly closest to maximum heating, the 0000 UTC sounding may be appropriate for U.S. adaptations. The adjustment is calculated by determining the mean cloud cover and geostrophic wind speed, both as a mean of values from evening, midnight, and morning. If the speed is 0-12 kt then

A=0, except if the cloud cover is 4-6 oktas (A=1.0) or 6-8 oktas (A=1.5). If the speed is 13-25 kt then A=0.5, unless cloud cover is 0-2 oktas (A=-1.5) or 2-4 oktas (A=0).

1.7. **Fog elimination prediction.** When insolation is present, clearance of fog depends on the thickness of the fog and the insolation available at a particular latitude and time of year.

1.7.1. JEFFERSON METHOD. The Jefferson method (Jefferson 1950) provides elaborate methods for modifying the sounding to illustrate the radiation inversion that will be present at dawn. Many of these concepts depend on British midnight radiosonde launches, and thus will not be discussed here. The key solution of the Jefferson method, however, is that where the dewpoint curve intersects the modified temperature trace (i.e. modified to show the deeper radiational inversion at dawn), the dewpoint line below this level is modified by extending it moist-adiabatically to the surface. This extended surface dewpoint represents the fog dissipation (clearance) temperature at which the fog will become stratus when the surface temperature heats to this value. By extending this surface dewpoint along the isotherms to the expected fog top level and then dry adiabatically to the surface, we obtain the surface temperature at which stratus will dissipate.

1.7.2. BARTHRAM METHOD. The Barthram method yields the time of fog dissipation (Barthram 1964) and was developed for British forecasting. Three values must be known: the depth of the fog in mb (hPa); T_1, the surface temperature at dawn in deg C; and T_2, the fog dissipation (clearance) temperature, which may be obtained from the Jefferson method above. The method is illustrated in Figure C8-1. Thin fog is considered to occur when visibility is 600 m or less, or has a depth of less than 20 mb. Each 300 ft of depth counts for 10 mb.

2. "Frontal" fog

Frontal fog will form when warm precipitation falls into a dry, cold air mass. The addition of moisture causes the air mass to rapidly reach saturation.

2.1. **Conditions.** Frontal fog is typically found in regions ahead of the warm front in well-developed baroclinic systems.

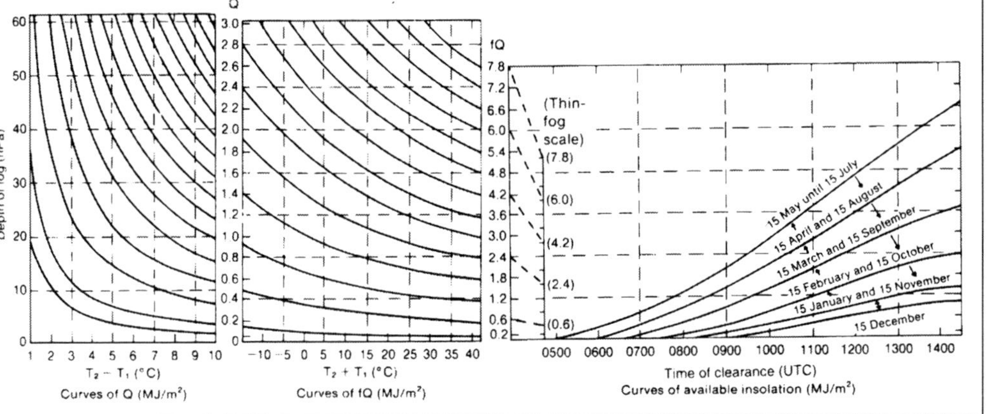

Figure C8-1. The Barthram fog clearance nomogram as developed for British forecasting. Starting at the left edge with fog depth, the forecaster proceeds horizontally right until the T_2-T_1 value is reached, then follows the curve to the right edge. The forecaster continues horizontally rightward across the border and into the next chart, and when the correct T_2+T_1 value is reached, the curve is followed to the right edge. In the same manner the middle chart is completed. The forecaster either bypasses the thin-fog scale or follows the entirety of the diagonal line, then proceeds horizontally rightward until the correct season curve is reached. The forecaster than proceeds straight down. Since this chart is for British use, UTC may be interpreted as local time.

2.2. **Prediction guidelines**.

2.2.1. HORIZONTAL. When the surface temperatures ahead of a warm front are *cooler* than the dewpoint temperatures behind it, then fog is likely. Cold ground is a contributing factor.

2.2.2. VERTICAL. Whenever the rain temperature exceeds the wet-bulb temperature of the cold air, fog or stratus will form. [CWFR]

3. Advection fog

Fog produced by air in motion or by transport of fog from one place to another is known as advection fog. True advection fog results when wind blows from a warm to a cold region, chilling the air and causing condensation to form. For example, when moist, warm air moves over colder ground or ocean, fog will result. However radiation fog may form over one locale, where the surface favors fog formation, and then be advected to another.

3.1. **Prediction guidelines**. When warm, moist air will advect over a cold surface, expect fog to form.

3.2. **Wind speed influences**. Light to moderate winds (0-13 kt) will have little to no effect on advection fog. Over the ocean, advection fog may exist at wind speeds of up to 25 kt since the lack of friction sustains a shallow chilled layer.

3.3. **Rules of thumb**.

- In nearly saturated air, light rain will often cause ground fog to form (fog less than 6 ft, or 2 m, in depth).
- Fog is enhanced in valleys due to cold air drainage from hill and mountain slopes. This effect may be pronounced in cold, stagnant high pressure areas, such as in southeast Idaho and northern Utah during the winter.

4. Sea fog

Fog is often formed over ocean or lake waters and then advected inland. The fog will dissipate as it crosses dry, unfavorable ground.

4.1. **Warm advection sea fog**. This type of fog occurs when a warm, humid air mass crosses a cold ocean or lake surface. The air mass loses heat to the water body, and is chilled to its dewpoint, causing fog and stratus to form. Localized cold spots in the water, such as those associated with upwelling, can cause the fog to form.

- If the water temperature is warmer than the air temperature of the air mass crossing above it, sea fog will occur.
- In the Gulf of Mexico, warm advection fog may occur in the wintertime when warm tropical air crosses over cold waters in the Gulf of Mexico, particularly when it reaches colder waters associated with ocean depths of less than 200 meters.

4.2. **Cold advection sea fog** ("steam fog", etc). A thin layer near the ocean surface is chilled to its dewpoint temperature, causing fog to form in this layer (giving the appearance of smoke coming from the water, i.e. "steam fog"). If the air mass dewpoint temperature exceeds the coldest water temperature it will cross, fog and fog banks may form.

- The depth of this fog is rarely greater than 120 ft; wind speed has little influence on the intensity of the fog and visibility can be zero even with a 30 kt wind [MT]. The duration of the fog is less than 18 hours, and areal coverage is not as widespread as with warm advection fog [MT].
- Sea fog and stratus are common in the Gulf of Mexico during the winter and early spring months. The heaviest fog is caused by cold arctic air masses reaching relatively warm Gulf waters.

5. Ice fog

Ice fog is a frozen fog caused by extremely cold temperatures and any modest amount of water vapor or pollutant. As such, it can be a pervasive weather type in urbanized areas of the Arctic.

5.1. **Formation**. Ice fog becomes a possibility when the temperature drops below -25 deg F (-32 deg C), and is likely in large towns and cities when the temperature falls below -39 deg C/F. [MT]

5.2. **Dissipation**. Studies have found that visibility rapidly improves when temperatures rise above -34 deg F (-37 deg C) [MT]. In the Alaskan interior in December and January, this will not occur during an extreme cold event until a major change in the weather pattern occurs, but in November, February, and March solar heating will allow this to occur by midafternoon [MT].

6. Tools

6.1. **Surface observations**. Surface reports are extremely useful, and trends in visibility, wind, and dewpoint depression can be directly tracked.

6.2. **Soundings**. Soundings have limited utility since they are widely scattered, available only every 12 hours, and do not adequately sample the delicate thermal configurations associated with fog.

6.3. **Satellite**. Visible imagery is ideally suited for detecting fog formation. Unfortunately, most fog develops at night. Infrared imagery is poorly suited for fog detection, since surface temperatures and fog temperatures are usually quite similar. A specialized satellite imagery product that uses the 3.9/10.7-micrometer channel difference product can be very useful and directly show the fog. Another technique involves examining the heat signatures of known lakes and noting when they become invisible; this works best for stratus layers and when animation is being used.

C9. FORECASTING

Haze

Haze is an atmospheric obscuration, originating from numerous pollution sources, that adds a dull grayish or bluish cast to distant objects.

1. Characteristics

1.1. **Cause**. Haze occurs chiefly due to the suspension of sulfates (SO_4^{2-}), which are produced from cloud-phase oxidation of sulfur dioxide (SO_2), which in turn originates largely from humans through the burning of fossil fuels and by steel and pulp/paper production (Ferman et al. 1981). Sulfur dioxide also occurs naturally from vegetative decay and volcanoes. The sulfate reaction occurs much more rapidly in the presence of humidity and water droplets, which is a reason why haze is linked to humid weather conditions. Other important contributions to haze come from volatile hydrocarbons released by various trees, nitrates, fine soil particles, and carbonaceous material from forest fires. Pollen and sea spray salt have sometimes been cited as haze contributors. *Recent research has downplayed a pervasive and longstanding association of maritime tropical air with haze events.*

1.2. **Horizontal extent**. The edge of haze is often sharply defined, and may be associated with fronts and surface boundaries. The top of haze is usually very well-defined and marks the top of the mixed boundary layer. The tops of cumulus clouds extend just above the haze layer.

1.3. **Vertical extent**. Haze is usually surface-based and typically stays confined within the boundary layer. This may give it a depth of hundreds of feet at night to many thousands of feet during the day as convective mixing progresses. A typical top of the haze layer is 8,000 ft during in the eastern United States during the warmest part of summer. The haze concentrates more strongly near the bottom part of the boundary layer at night and during the day mixing carries most of it aloft to the upper part of the boundary layer. As a result, by afternoon or evening it may unexpectedly affect the ability of aircraft pilots to see long distances.

1.4. **Intensity**. Haze appears thicker when the relative humidity is high, probably owing to the higher conversion rate of sulfur dioxide to sulfates, and possibly when the sun is at a lower angle in the sky.

1.5. **Arctic haze**. The phenomenon of Arctic haze was first documented in 1956 by M. Mitchell Jr., who described unusual haze bands of unknown origin in the high Arctic. No counterpart exists in Antarctica. Arctic haze is most frequent during anticyclone conditions in the spring months when mechanical turbulence is minimized. Trajectories originating *across* the Arctic basin have been found to be associated with Arctic haze in Alaska and Canada, sug-

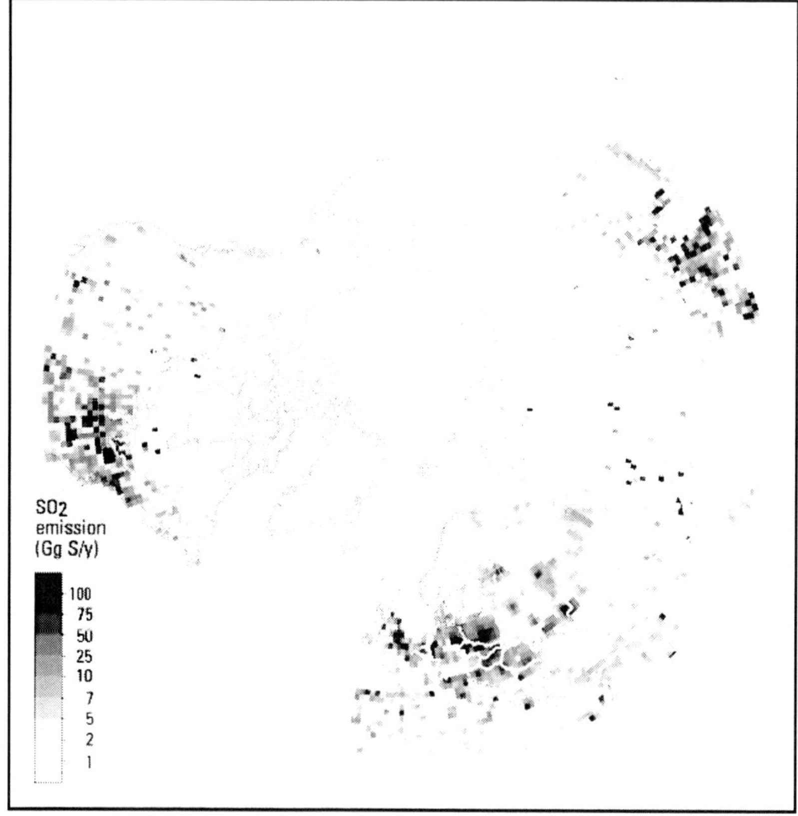

Figure C8-1. Anthropogenic sulfur dioxide (SO_2) emissions for 1985, from Rekacewicz (1998). Note the high concentrations in the United States "rust belt".

gesting advection of pollutants from Eurasia. Stagnant conditions during the polar night allow the pollutants to persist. (Shaw 1995)

2. Forecasting

2.1. **Source**. Regions of high sulfur dioxide emissions (Figure C8-1) include Ohio, Pennsylvania, and surrounding areas in the United States. The interior industrialized areas of central and eastern Europe are also significant sources of sulfur dioxide. Nitrate contributions to haze are chiefly from cities where vehicle traffic proliferates. Temporarily significant emissions of carbonaceous materials may arise from areas affected by wildfires, slash-and-burn agriculture, and volcanic eruptions.

2.2. **Development**. Conditions that favor the development of haze include:

- Slow low-level trajectories across pollutant source regions
- Weak or circular trajectories that allow pollutant-laden air parcels to stagnate or arrive multiple times
- Limited vertical mixing or subsidence (the lower and stronger the inversion, the greater the haze potential)

2.3. **Monitoring**. The surface winds and the 850 mb winds are the key charts for tracking the advection of haze. Trajectory forecasts, where available, can help determine the origin of air parcels that will arrive at the forecast area in the days ahead.

2.4. **Dispersion**. Haze is usually dispersed either through a combination of horizontal advection, vertical mixing, and/or continuous rainfall. Noteworthy mechanisms include:

- Widespread mesoscale convective systems which "overturn" the boundary layer across a large region (e.g. squall lines) will vent considerable amounts of haze into the upper troposphere, where they are dispersed. Individual thunderstorms are too small to have a measurable impact on haze.
- Arrival of a "tropical surge", noteworthy on the east U.S. coast, in which a shift in the Bermuda High (particularly a shift in its ridge axis past a forecast area) brings deep southerly flow and new trajectories from the relatively unpolluted subtropics (Corfidi 2004). These may occur not at all or several times per summer.

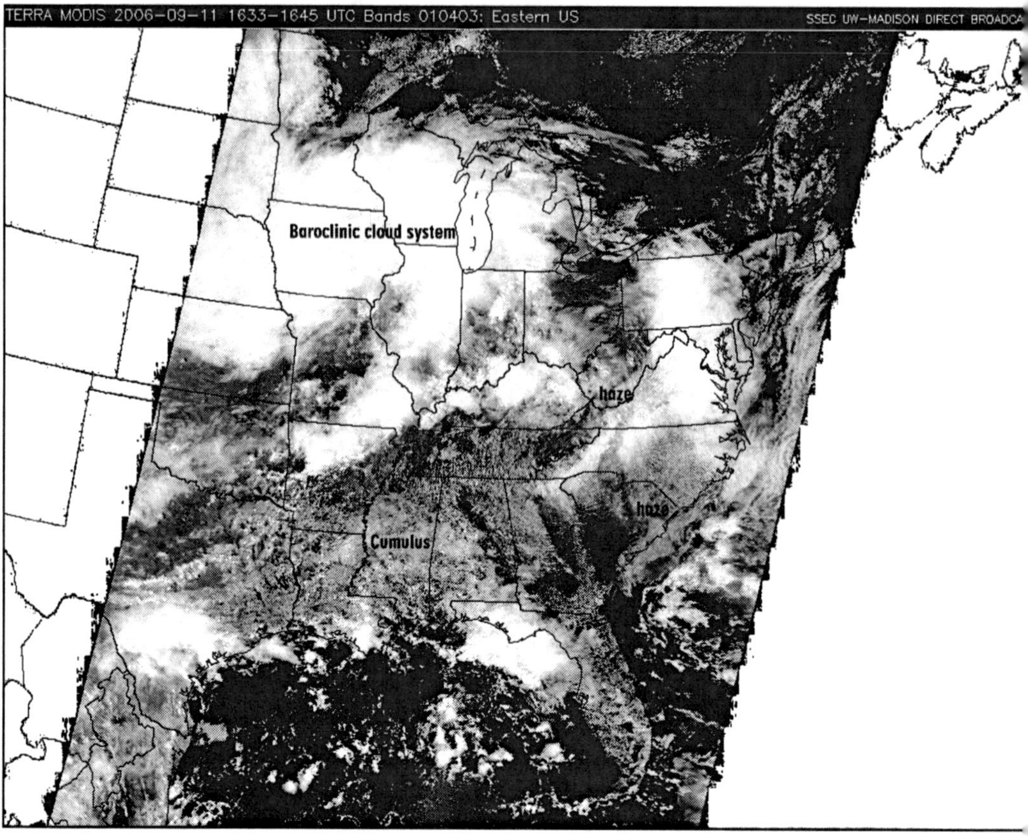

Figure C8-2. Haze. As shown above, MODIS TERRA satellite imagery (a good source is <http://rapidfire.sci.gsfc.nasa.gov/subsets/>) offers an excellent tool for tracking haze. Outstanding smoke and haze tracking information with imagery from MODIS and other systems are available at <http://alg.umbc.edu/usaq>.

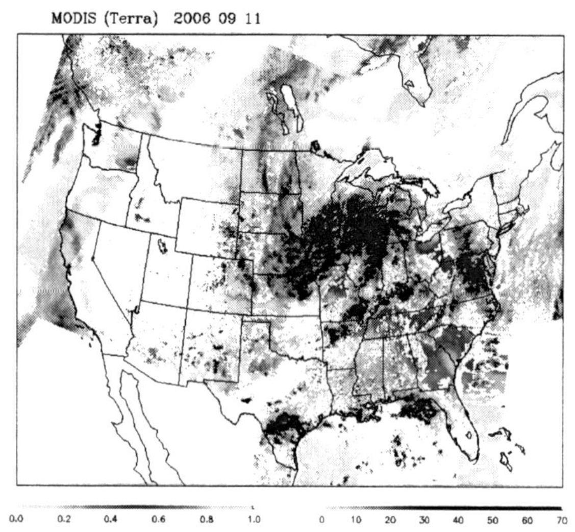

Figure C8-3. The MODIS AOD (Aerosol Optical Depth) product offers a unique way to measure atmospheric transparency in cloud-free areas (shown here by shading on periphery cloud patterns).

C10. FORECASTING

Temperature

Temperature forecasting is a challenge particularly when forecasting for coastal and mountainous communities.

1. Considerations

Temperature is influenced by the following factors:

- **Humidity** - Drier air allows a greater temperature variability.
- **Surface characteristics** - Urban areas, snow cover, etc.
- **Turbulent mixing** - Vertical transport of heat, winds, etc.
- **Advection** - Horizontal transport of heat.
- **Latent heat** - Widespread snowmelt requires additional heat.
- **Precipitation** - Wet-bulb cooling may take place.
- **Local effects** - Tertiary circulations, sea-land breezes, etc.
- **Air mass changes** - Frontal passages, etc.

2. Advection

Examine surface trajectories to find the origin of air parcels that will be in the forecast area during the forecast period. Determine the minimum or maximum temperature in that source area. Then apply adiabatic temperature corrections to compensate for the change in elevation (i.e. rising air cools at 5.5 F deg per 1000 ft of elevation, or 3.0 F deg if saturated; sinking air warms at 5.5 F deg per 1000 ft of elevation). Factor in any modification of the air mass, such as ground state, cloud cover, wind, and precipitation. [MT]

3. Wet bulbing

Evaporative cooling associated with the onset of precipitation will often rapidly change the surface temperature. The current wet-bulb temperature may be used as a first guess. Generally the air cools to the wet-bulb temperature after about 30 minutes of very heavy rain or 1 to 2 hours of light or moderate rain [FRB].

3.1. **Snow**. Falling snow gradually lowers the freezing level; however the reduction of the surface temperature to freezing is unlikely if the wet-bulb temperature of the surface is higher than 2.5 deg C in prolonged frontal precipitation, or higher than 3.5 deg C within extensive areas of moderate or strong precipitation associated with instability. [FRB]

3.2. **Thunderstorms**. In "non-frontal" thunderstorms the temperature of strong downdrafts reaching the ground are very close to the surface temperature of the saturated adiabat that runs through the intersection of the sounding's wet-bulb curve and the 0 deg C isotherm [FRB].

4. Minimum temperature

The most important tool in forecasting minimum temperature is assessing the type of air mass that will be over the area during the forecast period. When forecasting a cold front passage, the coldest minimum temperature should be expected on the second night after frontal passage [MT].

4.1. **Basic sounding method**. Follow the moist adiabat from the 850 mb dewpoint (700 mb if the station is above 850 mb) to the surface. If atmospheric changes are expected, use a forecast sounding instead. [MT] *Another layer besides 850 or 700 mb may be appropriate for a station depending on its elevation.*

4.2. **Dewpoint method**. Use the dewpoint at the time of maximum heating as a forecast minimum temperature for the following night. During non-summer periods on flat terrain or valley floors, and when skies are clear and winds are calm, the minimum temperature may be 4 to 7 deg F lower than this figure.

4.3. **McKenzie method** (*requires a predetermined constant*). To obtain the minimum air temperature, use the formula:

$$T_{min} = 0.5(T_{max} + T_d) - K$$

where T_{max} is the maximum temperature, T_{min} is the minimum temperature, T_d is the dewpoint temperature, and K is a local constant obtained through experimentation. Whether temperatures are in Celsius or Fahrenheit depends on the constant chosen. The K value typically depends on mean overnight cloud cover and wind speeds, thus a table should be constructed for the station. For example, in Birmingham, UK the K value for a Celsius computation varies from 0.5 on a windy night to 8.8 on a clear, calm night. Correction values also be devised for each month of the year. A 1990s Met Office study found that the McKenzie method is considered to be the best manual method for Great Britain forecasts.

4.4. **Barthrum method**. Used in the UK. Considered to be the most accurate for UK forecasting according to a Met Office 2002 study.

4.5. **Craddock method**. Originally developed to forecast overnight temperatures in the UK, the Craddock method uses a number of observed variables. The formula is as follows:

$$T_{min} = 0.316T_{12} + 0.548T_{d\,12} - 1.24 + K$$

where T_{12} is the midday air temperature in degrees Celsius and and $T_{d\,12}$ is the midday dewpoint in degrees Celsius. The value of K is shown in Table C10-1. *Note that all temperatures use Celsius.*

This method was considered to be the least accurate for the UK according to a Met Office 2002 study.

An update by Roger Lowe used by the U.S. Air Force is as follows:

$$T_{min} = 0.32T + 0.55T_d + 2.12 + C$$

where T and T_d are the 1200 UTC Celsius temperature and dewpoint [presumably since this is for the UK, other locations should use an early afternoon reading]. For C, if forecast winds overnight are 10 kt or less, C=-3 if cloud cover will be 0-2 oktas, C=-2 for 3 oktas, C=-1 for 4-5 oktas, and C=0 for 6-8 oktas. If forecast winds are 10 kt or more, C=0 unless 0-2 oktas are expected (C=-1) or 6-8 oktas are expected (C=+1).

Forecasting

Figure C10-1. Craddock & Pritchard minimum temperature method; this table computes a value for K given V_g (geostrophic wind speed in knots) and cloud cover in oktas.

	Cloud cover (oktas)			
V_g	0-2	2-4	4-6	6-8
0-12	5.0	5.0	4.0	4.0
13-25	4.0	4.0	3.0	2.0
26-38	3.5	3.0	2.5	2.5
39-52	2.5	2.5	2.5	3.0

5. Maximum temperature

Traditionally, maximum temperature forecasts in the U.S. have relied on MOS (model output statistics), and this is still very much the case. However there are some techniques that can augment or substitute for this method. The most important tool in forecasting maximum temperature is assessing the type of air mass that will be over the area during the forecast period.

5.1. **Simple sounding method**. In a clear to scattered sky, or when a warm front is approaching, follow the dry adiabat from the 850 mb (700 mb if station is above 850 mb) or 5,000 ft [AGL] temperature down to the surface. The resulting temperature is the estimate of maximum temperature. In broken to overcast skies, use the moist adiabat instead. One may also use the top of a nocturnal surface inversion, following its warmest temperature dry adiabatically to the surface; this method works best in clear to scattered skies in late spring or early autumn. [MT] *Another layer besides 850 or 700 mb may be appropriate for a station depending on its elevation. It possible to develop a lookup table for a given station that allows immediate conversion of temperatures aloft to forecast temperatures.*

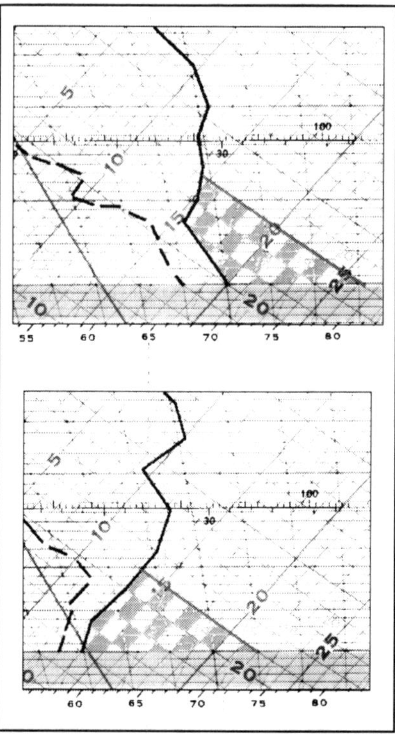

Figure C10-2. Graphical sounding method. This method assumes that, given the same amount of insolation and cloud cover, the atmosphere will heat up by a fixed number of "boxes" formed by the isotherms and dry adiabats. Here, the atmosphere heats up by about 25 boxes. In the top example, it will warm from 20°C to 26°C; in the bottom one, a warm lower troposphere allows a rapid increase from 14°C to 22°C.

5.2. **Graphical sounding method**. This method uses geometric area on the sounding (sometimes referred to as "box counting") and knowledge of the maximum temperature and geometric area on recent days. It assumes that, given a certain amount of solar radiation, the lowest part of the sounding will change by a certain amount of geometric area (see Figure C10-2), capped on the left by the environmental temperature and on the right by a selected dry adiabat, which displays the maximum temperature where it intersects the surface. This area can be related to the maximum temperature experienced on previous days under the same amount of cloud cover.

5.3. **Callen & Prescott method**. The technique uses the 1000-850 mb thickness as well as calendar month and observed cloud cover. It was developed for use in southern England. [FRB]

- STEP 1. Using known cloud cover and calendar month, use Table C10-3 to determine a temperature adjustment value.

Figure C10-3. Callen & Prescott method for determining temperature correction value, given month, cloud cover, and weather.

Class 0 - 3/8ths or less low and middle cloud; or
any fog confined to the dawn period
Class 1 - 4/8ths to 6/8ths total cloud cover;
or any fog clearing slowly during the morning
Class 2 - 6/8ths or more low and middle cloud;
or any fog clearing before noon
Class 3 - Predominantly overcast with precipitation
or persistent fog

Calculation of temperature correction (Fahrenheit degrees):

	CLOUD CLASS			
	0	1	2	3
JAN	-6.8	-7.5	-8.9	-10.1
FEB	-5.4	-6.7	-7.6	-8.9
MAR	-2.2	-3.8	-5.5	-6.9
APR	1.7	-0.3	-2.3	-4.3
MAY	4.2	2.2	0.1	-1.9
JUN	6.0	4.4	2.0	-0.2
JUL	6.8	4.7	2.3	0.2
AUG	5.4	3.5	1.5	-0.4
SEP	2.2	0.5	-0.8	-2.9
OCT	-1.6	-2.5	-3.8	-5.5
NOV	-4.4	-5.5	-6.7	-8.1
DEC	-6.7	-7.5	-8.7	-9.7

Forecasting

- STEP 2. Using the 1000-850 mb thickness, use Table C10-4 to determine an unadjusted maximum temperature. The formula $T_u = -192.65 + 0.156h$ may instead be used, where h is the thickness in meters.

- STEP 3. Add the unadjusted maximum temperature and temperature adjustment to find the predicted maximum temperature.

5.4. **Inglis method**. This technique uses the 1000-850 mb thickness and the calendar month to yield a first guess for maximum temperature. See Table C10-5 to determine the result. It assumes the pre-dawn 1000-850 mb lapse rate is 3/4ths that of moist adiabatic; the relative humidity of the layer is 75%, and the surface pressure is 1020 mb. It is not valid for days with extensive cloud cover. Since it was developed for British use it may have to be adjusted for other regions [FRB]

Table C10-4. Callen & Prescott unadjusted maximum temperature computation. Using 1000-850 mb thickness, read thousands, hundreds, and units of thickness in left columns, and units of thickness on top columns, and obtain value in degrees Celsius where they intersect.

THICKNESS (UNITS PLACE)

THK	0	1	2	3	4	5	6	7	8	9
1230	-0.8	-0.6	-0.5	-0.3	-0.1	0.0	0.2	0.3	0.5	0.6
1240	0.8	0.9	1.1	1.3	1.4	1.6	1.7	1.9	2.0	2.2
1250	2.3	2.5	2.7	2.8	3.0	3.1	3.3	3.4	3.6	3.8
1260	3.9	4.1	4.2	4.4	4.5	4.7	4.8	5.0	5.2	5.3
1270	5.5	5.6	5.8	5.9	6.1	6.2	6.4	6.6	6.7	6.9
1280	7.0	7.2	7.3	7.5	7.7	7.8	8.0	8.1	8.3	8.4
1290	8.6	8.7	8.9	9.1	9.2	9.4	9.5	9.7	9.8	10.0
1300	10.1	10.3	10.5	10.6	10.8	10.9	11.1	11.2	11.4	11.6
1310	11.7	11.9	12.0	12.2	12.3	12.5	12.6	12.8	13.0	13.1
1320	13.3	13.4	13.6	13.7	13.9	14.0	14.2	14.4	14.5	14.7
1330	14.8	15.0	15.1	15.3	15.5	15.6	15.8	15.9	16.1	16.2
1340	16.4	16.5	16.7	16.9	17.0	17.2	17.3	17.5	17.6	17.8
1350	17.9	18.1	18.3	18.4	18.6	18.7	18.9	19.0	19.2	19.4
1360	19.5	19.7	19.8	20.0	20.1	20.3	20.4	20.6	20.8	20.9
1370	21.1	21.2	21.4	21.5	21.7	21.8	22.0	22.2	22.3	22.5
1380	22.6	22.8	22.9	23.1	23.3	23.4	23.6	23.7	23.9	24.0
1390	24.2	24.3	24.5	24.7	24.8	25.0	25.1	25.3	25.4	25.6
1400	25.7	25.9	26.1	26.2	26.4	26.5	26.7	26.8	27.0	27.2
1410	27.3	27.5	27.6	27.8	27.9	28.1	28.2	28.4	28.6	28.7
1420	28.9	29.0	29.2	29.3	29.5	29.6	29.8	30.0	30.1	30.3
1430	30.4	30.6	30.7	30.9	31.1	31.2	31.4	31.5	31.7	31.8
1440	32.0	32.1	32.3	32.5	32.6	32.8	32.9	33.1	33.2	33.4

6. Regional techniques

6.1. **Texas**.

- ❑ With arctic air over north Texas that is less than two thousand feet thick, do not forecast more than 6 to 8 degrees of radiational cooling regardless of sky condition. [FTW]

6.2. **Europe**.

- ❑ In Europe, extreme cold is rare, but is usually associated with a westward extension of the Siberian high, and easterly flow.

Table C10-5. Inglis method for maximum temperature forecasting. Find intersection of 1000-850 mb thickness and calendar month, then read the maximum temperature in degrees Fahrenheit.

Thk	April	May	June	July	Aug	Sep
1280	44	46	46	46	45	43
1290	47	49	50	49	49	47
1300	50	52	53	53	52	50
1310	54	56	57	56	55	54
1320	57	59	60	60	59	57
1330	61	63	64	63	62	60
1340	64	66	67	67	66	64
1350	67	69	70	70	69	67
1360	71	73	73	73	72	70
1370	74	76	77	76	75	74
1380	77	79	80	80	79	77
1390	80	82	83	83	82	80
1400	84	86	87	86	85	84
1410	87	89	90	90	89	87
1420	91	93	94	93	92	91

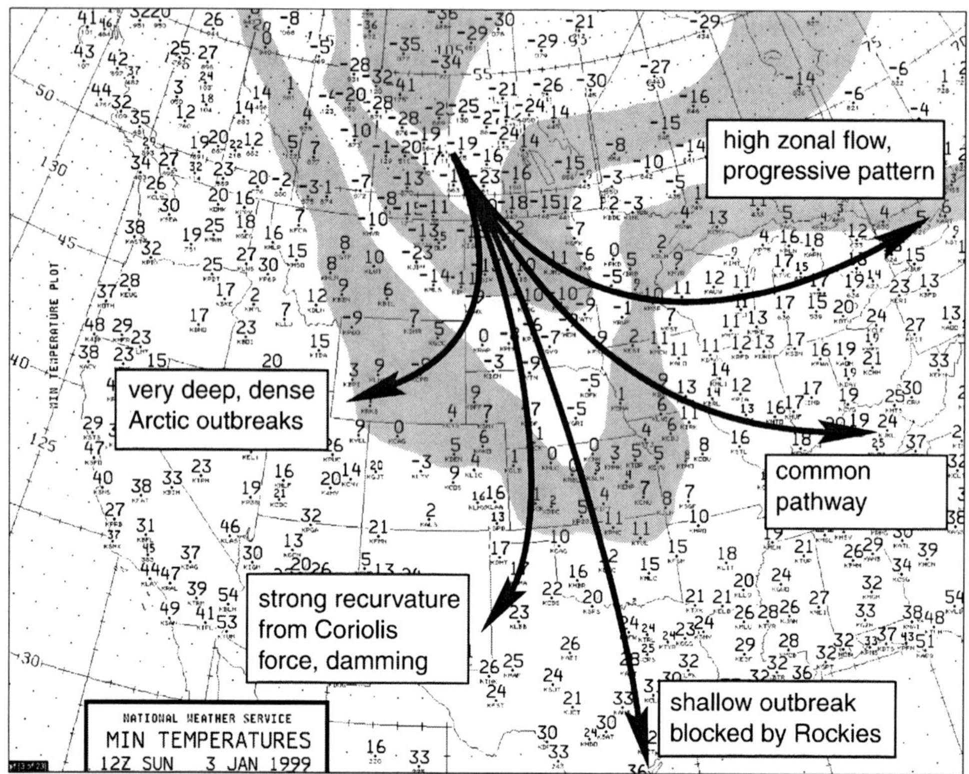

Figure C10-5. Polar outbreaks can be categorized by the paths taken by the air mass core. Curvature toward the left (towards the Great Lakes and Midwest) comprises the most common route, however, deviations to the right may occur if the air mass is sufficiently dense and deep. Very cold temperatures should not be forecast in southerly locales if the air mass core is expected to recurve strongly to the left.

C11. FORECASTING

Thunderstorms

The method of forecasting convective weather is a science and art in itself. The subject is so complex and multifaceted that space in this book does not allow it to be covered in detail. This section will focus only on key issues of thunderstorm probability, initiation, and location, and character, and will not expand on issues related to tornadoes, hail, and wind.

1. Fundamentals

1.1. **Prerequisites**. The following thunderstorm prerequisites were outlined by Doswell (1987):

- ❑ A moist layer of sufficient depth in the low or midtroposphere
- ❑ A steep enough lapse rate to allow for a substantial "positive area" on the sounding
- ❑ Sufficient lifting of a parcel from the moist layer to allow it to reach its level of free convection (LFC)

1.2. **Severe criteria**. Thunderstorms are considered to be severe according to the following definitions:

1.2.1. UNITED STATES. Winds of 50 kt (58 mph); hail ¾" in diameter, or tornadoes.

1.2.2. CANADA. Winds of 90 km/h; hail 2 cm in diameter; or rainfall of 50 mm within 1 hour or 75 mm within 2 hours.

1.2.3. GREAT BRITAIN. Winds of 90 km/h; hail 2 cm in diameter; a funnel cloud or tornado; or rainfall of 25 mm/h or more.

1.2.4. AUSTRALIA. Winds of 90 km/h; hail 2 cm in diameter; a tornado; or flash floods.

2. Thunderstorm types

There are several different "types" of thunderstorms which are recognized in meteorology. It must be noted that this is a conceptual framework. Storms rarely fit perfectly into these classifications and may share attributes with various other types.

2.1. **Unicell**. The unicell storm refers to a thundercloud that has only one primary updraft and downdraft. It is usually seen in benign, unstable environments, and may evolve into a multicell storm. Severe weather is usually minimal and the life cycle averages 30 to 60 minutes. A very strong unicell storm is often referred to as a *pulse thunderstorm*, and may contain small hail and high winds.

2.2. **Multicell**. A multicell consists of a thundercloud made up of numerous updraft and downdraft cells. Severe weather is generally confined to hail and high winds.

2.2.1. MULTICELL CLUSTER. A multicell cluster contains multiple cells that are chaotically organized. It is common in weakly-sheared environments with weak surface boundaries.

2.2.2. MULTICELL LINE. A multicell line has a linear structure, with old cells dying on one side of the line and new cells developing on the opposite side. It is sometimes seen in sheared environments and in association with strong surface boundaries.

2.3. **Supercell**. The supercell is a highly organized thunderstorm which features a continuous inflow/updraft state and a mesocyclone (an area of low pressure embedded within the updraft). Although most destructive storms are supercells, only 30% of supercells produce tornadoes. Supercells form in an environment with significant directional and speed shear with height.

2.3.1. LOW-PRECIPITATION (LP) SUPERCELL. The LP supercell features a poorly-defined downdraft; thus the storm is mostly represented by its updraft tower. Tornadoes are rare, and severe weather is usually limited to very large hail.

2.3.2. CLASSIC (CL) SUPERCELL. The CL supercell contains an active downdraft that cycles, periodically wrapping rain into the mesocyclone. The updraft and downdraft, and inflow and outflow, are equally dominant. The classic supercell may produce spectacular tornadoes.

2.3.3. HIGH-PRECIPITATION (HP) SUPERCELL. The HP supercell is dominated by downdraft areas, which often wrap opaque rain curtains into the mesocyclone. Tornadoes occur in some events and may be strongly rain-wrapped. It is a prolific producer of high winds and large hail.

2.4. **Squall line**. The squall line is a highly linear storm with a homogenous structure along its axis. A front-like downdraft area propagates the line forward, with updrafts and spectacular shelf (arcus) clouds occurring along the leading edge. True squall lines produce only rain and gusty winds, but at times may bring small- to medium-sized hail.

3. Forecasting

The proper use of all available tools is rarely more important than in thunderstorm forecasting. In particular, the use of the sounding (SKEW-T log p or tephigram) is absolutely essential. The process of forecasting thunderstorms consists of assessing the pattern, diagnosing environmental factors, predicting convective initiation, and anticipating thunderstorm behavior.

3.1. **Synoptic patterns**.

3.1.1. LARKO'S TRIANGLE. Larko's Triangle is a name given to the area with a vertex at the triple point of an extratropical cyclone, i.e. the intersection of the cold front and warm front, and extending about 100-200 nm into the warm sector. This region is where deepest moisture comes into closest proximity with favorable upper-level conditions. It works best during the cool season when storm development hinges strongly on the location of high theta-e parcels.

3.1.2. WARM FRONT. Since warm fronts typically develop in a southerly or southeasterly wind regime, the low-level flow often exhibits strong speed and directional shear when contrasted with winds aloft. As a result, there is an enhanced inclination for warm fronts to be associated with tornadic storms. The intersection of a low-level jet and outflow boundary has been noted as providing enhanced potential for development.

3.1.3. COLD FRONT. Advancing cold fronts may serve as sharp focusing mechanisms for thunderstorms. Due to the very linear nature of low-level convergence, squall lines are the favored mode of development; multicell and supercell storms are more common with slow-moving cold fronts.

3.1.4. OUTFLOW BOUNDARY. Outflow boundaries may persist 24 to 48 hours or more after convection disappears, and in some cases are resolvable only by close inspection of satellite, radar, and surface data. It is a common mode of initiation during the late spring and summer months. The inflow path and storm movement from one regime to another varies widely from situation to situation, so there are no clear-cut rules for outflow boundary convection. The intersection of a low-level jet and outflow boundary has been noted as providing enhanced potential for development.

3.1.5. DRYLINE. In the Great Plains, since the dryline marks the westward limits of tropical moisture, it serves as the first location where upper-level destabilization reaches an area with buoyant parcels. The author has noted that storms often fail to initiate along a dryline if there is insufficient convergence in the wind field. Since the 1970s there has been some interest in "dryline bulges", where part of the dryline accelerates forward; it has been postulated that there is enhanced probability for severe weather along or north of this bulge, presumably due to enhanced surface convergence and enhanced upper-level dynamics.

3.1.6. NEAR A SURFACE LOW. In the absence of fronts and drylines, storms may form in the convergence surrounding a surface low, especially where it is focused such as along troughs. The best directional shears, which support the greatest likelihood of rotating updrafts, are usually located northeast of the surface low.

3.1.7. OROGRAPHIC SOURCES. In a moist, unstable atmosphere, prominent topographic features may serve as sources for mechanical lift and as a mechanism to advect moisture quickly to higher terrain (boosting parcel theta-e). In the Great Plains, noteworthy features include the Raton Mesa, the Palmer Divide, the Black Hills, and the mesas at the Colorado-Wyoming-Nebraska border.

3.1.8. COLD-CORE LOWS. Cold-core upper-level lows, such as those in the wake of frontal systems, may be associated with steep lapse rates and favor the development of showers and thunderstorms. These tend to be strongly diurnal and dissipate rapidly after sunset. Wind shear profiles depend on whether the system is vertically stacked (barotropic) or tilted (baroclinic), but in most cases do not favor the development of rotating updrafts.

3.1.9. TROPICAL CYCLONES. Thunderstorms are common within tropical cyclones. The highly-sheared environments may lead to the development of tornadoes, especially in the forward right quadrant (forward left in the southern hemisphere).

3.2. **Environmental factors**. These largely relate to tangible quantities which allow for thunderstorms and associated severe weather to develop: instability and shear. Key tools include soundings and surface charts.

3.2.1. INSTABILITY. Instability is tied to updraft speed, which in turn is tied to storm severity.

❑ *Positive energy.* This is the amount of positive area on the sounding (where a lifted parcel is warmer than the environment), not counting capped layers. CAPE (see Stability Indices section) is by far the most reliable indicator of instability. All other indices should be regarded as quick estimations. *Soundings must be modified* to reflect expected parcel and environmental conditions at convection time.

❑ *Negative energy.* This is the amount of negative area on the sounding (where a lifted parcel is colder than the environment), capping a parcel from deep convection. The CINH expression (see Stability Indices section) is equivalent to "negative CAPE" and uses the same units as CAPE. The area above the equilibrium level is not counted. Values of CINH 200 j/kg or more are associated with strong capping.

3.2.2. SHEAR. There are three basic types of shear expressions used in storm forecasting:

❑ *Deep-layer shear* is very basic quantity describing the vector difference between low-level and upper-level winds. High values in general are associated with updraft-downdraft separation and longer-lived storms. The most common measure of deep-layer shear is the 0-6 km shear vector (BRN Shear); a shear of 50 kt has been cited as a threshold for supercells.

❑ *Anvil-level storm-relative winds* gives a close measure of the downdraft-updraft separation that can be expected. It is measured directly between the storm movement vector and the 9-11 km (~300 mb) wind vector in knots. In a supercell environment values of more than 60 kt are associated with LP (low-precipitation) structures; values of less than 40 kt are associated with HP (high-precipitation) structures.

❑ *Helicity* is usually assessed in the lowest 1 to 3 km of the atmosphere. It is most useful when evaluated relative to the storm's movement, i.e. *storm relative helicity* (SRH). High values of SRH contribute to greater rotation in updrafts.

3.2.3. COMBINED INSTABILITY AND SHEAR. Instability and shear must be evaluated in tandem, because high instability without shear produces pulselike nontornadic storms while low instability with high shear results in no storms or a lack of initiation due to shearing apart of cumulus towers.

❑ *Energy-Helicity Index (EHI)* is a useful predictor that combines CAPE and Storm Relative Helicity. It is most useful for tornado prediction. Its skill at predicting thunderstorm initiation and general severe risk thresholds is limited because it only examines low-level helicity and does not assess deep-layer shear.

❑ *Vorticity Generation Potential (VGP)* is similar to EHI, representing a combination of CAPE and shear. However in VGP the square root of CAPE is used, diminishing its weight, and the 0 to 3 km shear vector is used. Supercell tornadoes are rare with a VGP of less than 0.3 m/s^2 but are likely above 0.6.

3.3. **Initiation**. New thunderstorms will form wherever a parcel is able to ascend through a deep, unstable layer of the atmosphere. In many cases storms may be suppressed by an inversion (such as a the ubiquitous "cap" on the Great Plains) until the inversion is weakened. Thunderstorms are not "triggered" by boundaries and upper dynamics; rather these factors make parcel ascent more likely in a particular location by destabilizing the atmosphere through large-scale lift (which in turn may be caused by convergence or divergence).

3.3.1. GENERAL LOCATION. Given homogenous upper-level conditions over an area, initiation is most likely where there is the highest theta-e (equivalent potential temperature) of parcels. Given homogenous low-level conditions over an area, initiation is most likely where there is the weakest convective inhibition (cap, etc) aloft.

3.3.2. BOUNDARIES. Due to their role in convergence of the low-level wind field, all boundaries in a warm, moist airmass should be examined as potential initiation areas for thunderstorms.

3.3.3. MOISTURE. Parcel moisture (dewpoint) is about two to four times as important as temperature in assessing whether a cap will break. A four-degree drop in surface temperature can be compensated by a one- to two-degree increase in dewpoint.

3.3.4. ELEVATION. Given a parcel at a high elevation and a low elevation with the same dewpoint, initation is more probable in higher terrain. Theta-e analysis can remove the effect of elevation; higher theta-e values correspond to warmer parcel temperature.

3.3.5. TORNADIC AREAS. Tornadic storms are most favored where there is (a) relatively isolated development of cells and (b) where winds in the lowest 1-2 km of the atmosphere provide the greatest amount of directional shear. Typically this is east and poleward of a surface low, where low-level winds contrast most sharply with winds aloft.

3.3.6. UPPER LIFT. Sources of upper lift that destabilize the atmosphere (jet streaks, short waves, etc) must be carefully tracked. The use of hourly isallobaric fields (pressure change) can be valuable in tracking these upper-level disturbances. Use the altimeter setting (QNH), not sea-level pressure, as the latter is corrected with 12-hour temperature means. The *movement* of patterns is more important for tracking disturbances than the actual values, which may be influenced by diurnal pressure tides, localized heating, and so forth. However, any persistent pressure fall or pressure rise will usually cause a response in the surface wind field, bringing effects on shear and convergence.

3.4. **Character**. Assuming that storms develop, it's important to anticipate the type of storm behavior that can be expected. This can be determined from storm motion, shear profiles, and instability.

3.4.1. STORM MOTION. The storm motion is important information for not only determining tracks but in assessing whether storms that develop in a particular area will move into favorable or unfavorable environments. There are a few different techniques for predicting storm motion:

❑ *Quick normal cell motion*: Using a hodograph (or using vector calculation), find the average of all winds in the troposphere. This yields a *mean sounding wind speed* and provides an initial guess at nonsevere storm motion.

❑ *30R75 deviant motion rule.* Deviant right motion is 30 degrees to the right and 75% of the mean sounding wind speed. Improved on by the Bunkers method. (Maddox 1976)

❑ *Bunkers deviant motion rule*: Using a hodograph, find the average of the 0-0.5 km and 5.5-6.0 km winds. Connect these two averages. Construct a line perpendicular to this vector that intersects the normal cell motion, *N*. The deviant right motion is 15 kt to the right of *N* and deviant left motion is 15 kt to the left of *N* along this perpendicular line. (Bunkers et al 2000)

3.4.2. STORM-RELATIVE ANVIL WINDS. The speed of anvil-level winds (200-300 mb) relative to the storm motion largely determines the type of precipitation structure that will develop. Speeds of 60 kt or greater favor significant separation of updraft and downdraft (LP supercell structures), while those of 35 kt or less favor significant precipitation within the storm (HP supercell structures).

3.4.3. BOUNDARY ALIGNMENT. The surface boundary alignment relative to the mean tropospheric flow is a key factor. Alignment parallel to the upper-level winds favors long-lived storms but introduces the possibility of seeding (contamination of precipitation into downstream updrafts). Boundary alignment perpendicular to the mean tropospheric flow minimizes seeding but introduces the possibility of storms departing the boundary into an unfavorable air mass.

3.4.4. MESOSCALE CONVECTIVE SYSTEMS (MCS). Large areas of organized thunderstorms (MCS areas) are favored along strong, linear boundaries, such as strong outflow boundaries and cold fronts. There is often greater competition between cells and a reduced threat of tornadic activity.

C12. FORECASTING

Tropical cyclones

1. Fundamentals

1.1. **Requirements**. A tropical cyclone requires all of the following to achieve self-sustaining development:

1.1.1. HEAT SOURCE. Warm ocean waters in excess of 80°F with about 200 ft or greater depth are required.

1.1.2. AN UNSTABLE ATMOSPHERE. This allows thunderstorms: the catalyst for the release of latent heat.

1.1.3. CORIOLIS FORCE. The tropical system must be at least 300 nm (5 deg of latitude) from the equator to bring the system's wind fields into balance.

1.1.4. WEAK VERTICAL SHEAR. A vector difference of less than 20 kt from the bottom to the top of the tropopause is essential to prevent disruption of the system's wind fields.

1.1.5. A PRE-EXISTING DISTURBANCE. An easterly wave is the most common type of disturbance.

1.2. **Descriptors**. The following names have been standardized to describe tropical weather systems:

1.2.1. TROPICAL DEPRESSION. Organized system of clouds and thunderstorms with maximum sustained winds of less than 34 kt (39 mph).

1.2.2. TROPICAL STORM. Maximum sustained winds of 34 to 63 kt (39 to 73 mph).

1.2.3. HURRICANE, TYPHOON, or TROPICAL CYCLONE. Maximum sustained winds of 64 kt (74 mph) or greater.

1.2.4. SUPERTYPHOON. West Pacific basin tropical cyclone with maximum sustained winds of 130 kt (150 mph).

1.3. **Saffir-Simpson scale**. The Saffir-Simpson intensity scale is the leading scale in use throughout the American region.

1.3.1. CATEGORY 1. Winds of (74 to 95 mph). Damage is primarily to unanchored mobile homes, shrubbery, and trees.

1.3.2. CATEGORY 2. Winds of (96 to 110 mph). Roofing, door, and window damage. Piers and small craft damaged by flooding.

1.3.3. CATEGORY 3. Winds of (111 to 130 mph). Structural damage to small residences. Mobile homes destroyed.

1.3.4. CATEGORY 4. Winds of (131 to 155 mph). Complete roof failure on small residences. Major erosion of beach areas.

1.3.5. CATEGORY 5. Winds of (156 mph) or more. Complete roof failures on many buildings, with some buildings destroyed.

2. Analysis

Analysis is traditionally performed using satellite and surface observations, but other tools such as radar and dropsonde have fallen into widespread use.

2.1. **Fix positioning**. Highly accurate positioning is critical for forecasting. Visible imagery should be chosen over infrared, since infrared imagery's sensitivity to cold clouds may provide a false fix in the mid-levels of the storm. The Dvorak technique is the preferred method for fixing a storm using satellite. If available, radar and surface data can provide accurate fix estimates. If no circulation is noted, the storm is placed at the center of the most persistent convection during the past 24 hours.

2.2. **Dvorak Intensity Technique**. The Dvorak technique was developed in the 1970s to allow accurate estimates of tropical cyclone intensity from satellite observations. It examines cloud features and characteristics. *As of press time, the 1984 Dvorak intensity technique could not be located for inclusion in this book.*

3. Forecasting

Forecasting should take into consideration persistence, climatology, and numerical methods.

3.1. **Influences on movement**. Tropical cyclone movement is influenced by interactions with...

- ... westward extensions of the subtropical ridge
- ... north-south migrations of the subtropical ridge axis
- ... points of weakness in the subtropical ridge
- ... mid-latitude extratropical systems
- ... other tropical systems
- ... weak flow regions where steering currents are poorly defined

3.2. **Steering methods**. Tropical cyclones will tend to move . . .

- . . . with the average deep-layer wind field within which they are embedded. However the storms do change this environment with time. This layer extends from 850 to 500 mb for 990-999 mb storms; 850 to 400 mb for 970-989 mb storms; 850 to 300 mb for 950-969 mb storms; 850 to 250 mb for 940-949 mb storms, and 700 to 200 mb for deeper storms (Velden 1990).
- . . . with an estimated steering current, as a first guess. For most storms this is at about 300 mb. Weaker storms may steer with the 400-500 mb winds, and stronger ones with the flow up to 100 mb. *(Empirical)*
- . . . toward ridges at 200-400 mb.
- . . . toward the maximum surface pressure falls.
- . . . less westerly with increasing latitude (recurvature).

Forecasting

3.3. **Recurvature**. Tropical cyclones will recurve as they move away from the equator; that is, they will move less westerly as they impinge on the prevailing westerlies.

3.3.1. RECURVATURE MODES. Recurvature is most likely when a large-amplitude upper-level trough, extending southward from the westerlies, is located within a few hundred miles to the west of the storm center; it is also likely with well-marked low-latitude troughs building northward into the westerlies and weak troughs between two separate high cells (George and Gray 1977). When the equatorward edge of an approaching major mid-latitude trough is further equatorward than the tropical cyclone, expect recurvature (Riehl and Shafer 1944).

Recurvature is also related to the strength of the 300 mb winds poleward of the storm (Burroughs and Brand 1972).

3.3.2. NONRECURVATURE MODES. The presence of a strong subtropical (barotropic) high poleward of a tropical cyclone makes recurvature highly unlikely. A weak subtropical high poleward of a tropical cyclone, with strong zonal upper-level flow, also imhibits recurvature. (George and Gray 1977)

3.3.3. OBJECTIVE DISCRIMINATORS. One rule of thumb used through the 1960s and 1970s, before emphasis was placed on upper-tropospheric wind fields, uses the 850-500 mb thickness value 500 miles (7.5 deg) northwest (southwest in the southern hemisphere) from the storm center; values of *4270 m or lower* suggest imminent recurvature; values of *4330 m or more* suggest no imminent recurvature.

3.4. **Intensity changes**. The change in the intensity of the storm can be assessed by monitoring trends on infrared satellite imagery where cloud tops are colder than -65°C. If this area expands, deepening of the storm can be expected. If it shrinks, weakening can be expected.

3.4.1. ANTICIPATING INTENSITY CHANGES. The following indicators are suggestive of intensification:

- ❑ Subtropical ridge is further north than usual, allowing for deeper tropical easterlies
- ❑ Storm movement is less than 13 kt
- ❑ The system has a poleward component of motion; this takes it to latitudes where Coriolis effect increases, inhibiting filling
- ❑ A migratory anticyclone passes poleward of the storm's center; this increases pressure gradients
- ❑ The storm passes underneath an upper trough with relative motion between the two
- ❑ Sea surface temperatures 79°F or more over deep waters
- ❑ Slow progression of long wave pattern

3.4.2. EYEWALL REPLACEMENT CYCLES. Eyewall replacement cycles cause weakening and strengthening cycles on the order of hours. As a tropical cyclone becomes very strong (winds of above 100 kt) its eyewall contracts to less than 15 miles in diameter. At this point the outer rainbands form a new eyewall which replaces the inner eyewall. During this phase the storm weakens and the central pressure increases. Once the inner eyewall is replaced, the storm deepens again.

3.5. **Tornadoes**. Though small-scale vortices are common in the eyewall of strong hurricanes, strong shear profiles near the storm may allow for organized, rotating updrafts in its peripheral rainbands. Early hurricane studies found that the right front quadrant of the hurricane, especially at a distance of 100-150 nm from the eye, was a favored location for tornadic activity. Signals for possible tornadic activity include

- ☐ Tropical cyclones intensifying just before landfall
- ☐ Rapid filling of the storm center after landfall
- ☐ Rapid cooling of the storm center on infrared imagery
- ☐ Wind shear of 40 kt or more between surface and 850 mb

Figure C12-1. Dvorak intensity numbers and their associated wind speeds, sea level pressures, and Saffir-Simpson intensities.

T Number	Wind Speed kt	mph	Sea-level Press. Atlantic	NW Pac	Saffir-Simpson Cat (approx)
1	25	29			
1.5	25	29			
2	30	35	1009 mb	1000 mb	
2.5	35	40	1005 mb	997 mb	
3	45	52	1000 mb	991 mb	
3.5	55	63	994 mb	984 mb	
4	65	75	987 mb	976 mb	1
4.5	77	89	979 mb	966 mb	1-2
5	90	104	970 mb	954 mb	2-3
5.5	102	117	960 mb	941 mb	3
6	115	132	948 mb	927 mb	4
6.5	127	146	935 mb	914 mb	4
7	140	161	921 mb	898 mb	5
7.5	155	178	906 mb	879 mb	5
8	170	196	890 mb	858 mb	5

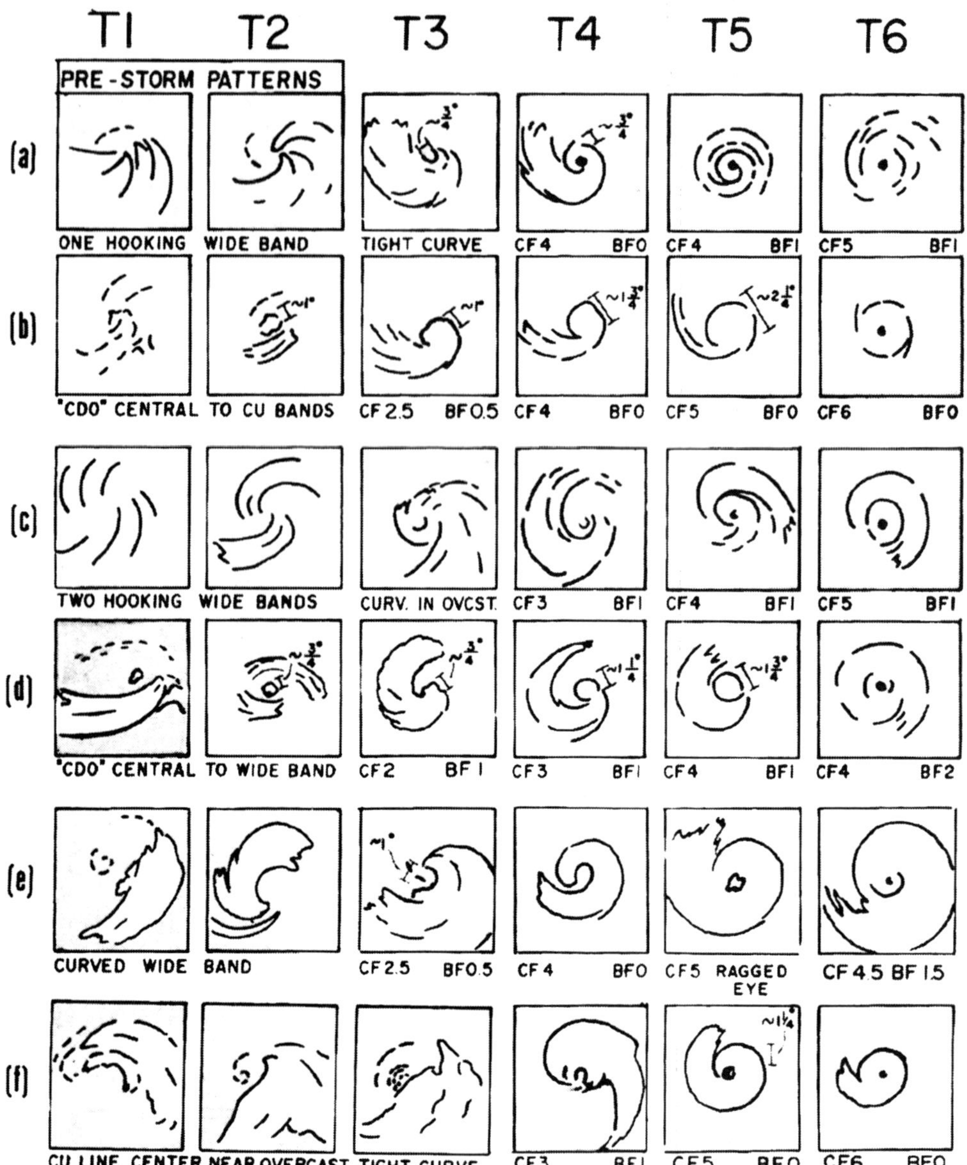

Figure C12-2. Dvorak method cloud characteristics. This chart may be used as a reference for cloud patterns associated with various tropical cyclone intensities. However it cannot be used by itself to obtain an intensity value.

C13. FORECASTING

Winter weather

1. Determining type

The mid-level is assumed to be the layer from 1500 to about 4000 ft, or within the cloud producing the precipitation. The low-level is that area beneath the mid-level.

1.1. **Thermal structure**. It is imperative that the forecaster examine the structure of temperature in the lowest 2 km of the atmosphere. This has great bearing on the type of winter weather that might occur. In the paragraphs below, "surface" means the

Figure C13-1. Quick reference guide for forecasting winter precipitation type. The term "warm" is above freezing; "cold" is below freezing.

STRUCTURE	RESULTING PRECIPITATION
Solid precipitation remains in cold layer or falls through warm layer of less than 600 ft depth	Solid (snow)
Solid precipitation falls through warm layer of 600 ft depth or more	Solid/liquid mix (rain/snow mix)
Solid precipitation falls through warm layer of 1200 ft depth or more	Liquid (rain)
Liquid precipitation passes through warm layer, or passes through cold layer of less than 800 ft depth	Liquid (rain)
Liquid precipitation passes through cold layer of 800 ft depth or more	Frozen liquid (freezing rain)

layer below about 500 meters, while "aloft" means the layer just above. "Cold" means below freezing and "warm" means above freezing.

1.1.1. WARM SURFACE, WARM ALOFT. This temperature profile is associated with tropical and warm-season precipitation. It will always result in rain since no part of the profile is below freezing.

1.1.2. WARM SURFACE, COLD ALOFT. Associated with unstable (often convective) conditions, such as those behind a cold front. Such patterns usually produce snow which reaches the ground as snow or melts as rain. Which one results depends upon the *depth* of the warm layer. Studies have found that *rain will occur if the warm layer is 750 to 1500 ft deep*, depending on snowflake type, drop size, wet bulb temperature, and lapse

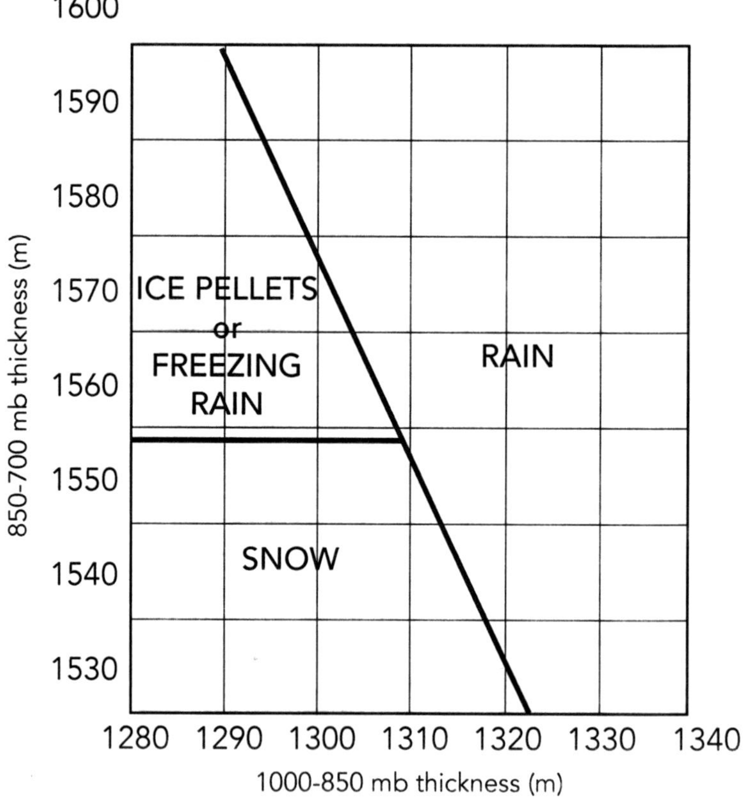

Figure C13-2. Precipitation type graph based on thickness, valid for areas below 500 ft MSL.

rate. One study found that the chance of snow is 50% when the depth of the warm layer is 920 ft.

1.1.3. COLD SURFACE, WARM ALOFT. Associated with warm fronts and "overrunning" situations. This presents the most difficult type of winter forecasting situation. The questions are: *(a) How much melting will occur in the warm layer, and (b) how much refreezing will occur in the cold layer beneath. These will be treated separately:*

❑ *Elevated warm layer*. If the elevated warm layer is sufficiently warm, all precipitation will be melted before falling into the cold layer beneath. One rule of thumb states that if the warm layer is greater than 600 ft deep, or the maximum temperature in the elevated warm layer exceeds 3-4 °C, then complete melting will occur.

❑ *Cold surface layer*. When entering a cold layer, all frozen precipitation will arrive intact (i.e. any snow will impact as snow). If this layer is sufficiently cold and deep, liquid precipitation will freeze as it falls and impact as ice pellets (sleet). Otherwise liquid precipitation will not freeze impact as rain (which will freeze into freezing rain if the ground and air is cold enough). One rule of thumb states that if the cold layer is greater than 800 ft deep, then liquid drops will be frozen by the time they impact.

1.1.4. COLD SURFACE, COLD ALOFT. This profile is associated with deep winter cold and polar conditions. Snow may always be expected.

Forecasting

1.2. **Rules of thumb.**

1.2.1. WARM SURFACE TEMPERATURES. When the surface temperature is 39 °F or greater, precipitation will usually be confined to liquid or mixed precipitation.

1.2.2. RELATION TO THICKNESS ISOPLETHS. The 1000-500 mb thickness 540 dam line is often used as a discriminator between rain and snow. This only works for stations near sea level.

1.2.3. RELATION TO 500 MB TEMPERATURE. The 500 mb -20 °C line is often used as a discriminator between rain and snow. This only works for stations near sea level.

1.2.4. SNOW INDEX. The snow index is:

$$S = (H_{850} - H_{700}) + 2 \times (H_{1000} - H_{850})$$

where H is the geopotential height at the levels shown in meters. A value of greater than 4179 indicates liquid precipitation; 4179 indicates mixed precipitation, and less than 4179 indicates solid precipitation.

2. Determining snow accumulation

2.1. **Using observations**. The use of weather observations may be affected by fog and blowing snow.

- For light snow, with visibility 5/8 mile or greater, accumulation is less than 0.2 in/hr.
- For moderate snow, with visibility 5/16 to less than 5/8 mile, accumulation is betwen 0.2 and 1.6 in/hr.
- For heavy snow, with visibility less than 5/16 mile, accumulation is 1.6 in/hr or more.

3. Winter weather patterns

3.1. **Extratropical cyclone**. Heavy snow usually occurs in the northwest quadrant of the surface low. The duration averages 12 to 36 hours.

- Snow is not likely to the right of the 850 mb low track.
- Heavy snow will usually fall in a band 100-200 miles wide to the *left* of the track of the surface low. This may also be 75-250 miles to the left of the 850 mb surface low track.
- The axis of heavy snow is typically 150 nm to the left of the vorticity maximum track or the 534 dam 1000-500 mb thickness line.
- A longtime rule of thumb states that precipitation will begin as the 700 mb ridge passes, and end as the 700 mb trough passes. Heavy precpitation will begin as the 500 mb ridge passes, and end as the 500 mb inflection point arrives.

3.2. **Sharp cold front passage**. A band of snow may lie behind a cold front, sometimes within a deep trough. The duration of this snow is usually less than 4 hours.

- Low dewpoint depressions (6 °C or less) should exist at 850 and 700 mb.
- Cyclonic flow exists at 700 and 500 mb.
- The heaviest snow tends to exist in the region between the 850 mb and 700 mb trough.

3.3. **Lake-effect snow**.

- The lake water should be at least 13 °C warmer than the 850 mb temperature.
- The lake water should be at least 20 °C warmer than the 700 mb temperature.
- Cold advection should be indicated at 700 and 850 mb.
- Do not forecast lake-effect snow with moderate or strong warm air advection.
- Wind speeds should be moderate. Strong winds (less for small lakes, more for large lakes) will advect the air mass too quickly to result in destabilization.

4. Regional

4.1. **Alaska**.

4.1.1. FAIRBANKS usually receives winter snowfall when occluded fronts approach from the south or southwest. The southwest quadrant of Fairbanks is open to the sea and is free of high mountain barriers, allowing moist maritime flow into the area. A south wind will rarely give Fairbanks any weather due to blocking from the Alaskan Range. (Weber, 1980)

4.1.2. ANCHORAGE receives its most significant winter precipitation when southerly flow exists at 500 mb. The critical location for the main 500 mb low center is in the Bering Straits, particularly if cyclonic vorticity is located upstream and a surface low is in the Gulf of Alaska. (Weber, 1980)

4.2. **Conterminous United States**.

4.2.1. EASTERN COLORADO receives heavy snowfall associated with cold air damming in anticyclonic upslope conditions, especially when southeasterly 700 mb winds are possible. Mesonets will show a small-scale quasistationary cyclonic shear boundary aligned parallel to and east of the Front Range, with heaviest snow just to the west of this boundary. (Dunn, 1987)

C14. FORECASTING

Stability indices

This section is a summary of stability indices used worldwide, with relevant convective forecasting indices for shear also included. *Stability indices should never substitute for a forecaster's own analysis and diagnosis process.* Their use operationally should be limited to a select few of the most relevant values for a given situation.

1. Lapse rate

The lapse rate is simply the rate of decrease in temperature through a given height. It is typically expressed in degrees Celsius per kilometer, with values below 5.5 C°/km corresponding to stable air and values above 9.8 C°/km corresponding to absolutely unstable air. A more useful expression of lapse rate for convective forecasting is the 700-500 mb temperature difference (i.e. "700-500 mb delta-T"):

$$dT = T_{700} - T_{500}$$

It is useful for assessing how the synoptic-scale environment contributes to CAPE assuming constant parcel theta-e. Rough meanings are:

14: Marginally unstable (warm season)
17: Marginally unstable (cold season)
20: Moderately unstable
26: Extremely unstable (more than about 26 is impossible)

2. Convective Available Potential Energy (CAPE)

CAPE is currently the most widely used predictor for both thunderstorm potential and severe weather risk. It was defined in 1982 by Morris Weisman and Joseph Klemp. The form of the equation is shown here:

$$\text{CAPE} = \left(\sum_{LFC}^{EL} \left[\frac{(T_p - T_e)}{T_e} g \right] \right) \Delta z$$

which yields integrated instability in joules per kilogram. The lifted parcel can be from the surface (SBCAPE), from a mixed layer

(MLCAPE), or from a level that yields the most unstable profile (MUCAPE).

Typical values are:
<300: Mostly stable, little or no convection
300-1000: Marginally unstable; weak thunderstorm activity
1000-2500: Moderately unstable; possible severe thunderstorms
2500-3500: Very unstable; severe thunderstorms; possible tornadoes
3500+: Extremely unstable; severe thunderstorms; tornadoes likely

3. Convective Inhibition (CINH)

Convective Inhibition is calculated in the same manner as CAPE except for areas along the parcel lift where the parcel is colder than the surrounding air. In effect, it figures areas that are negatively buoyant. It was developed in 1984 by Frank Colby.

Typical values are:
<0: No cap
0 to 20: Weak capping
21-50: Moderate capping
51-99: Strong capping
100+: Intense cap; storms not likely

4. Downdraft Convective Available Potential Energy (DCAPE)

This is the region on the sounding between the trace of a parcel descending wet-adiabatically and that of the environment. The parcel should originate from the minimum wet-bulb temperature in the 500-700 mb layer and be sunk wet-adiabatically to the ground. Another method originates the parcel at the wet-bulb temperature of the 600 mb level. The geometric area between this wet adiabat and that of the warmer environmental trace to the right establishes the DCAPE. This index indicates the ability of a storm to produce fast-moving downdrafts and strong outflow. It is suggested that higher DCAPE values are associated with stronger rear-flank downdrafts in supercell thunderstorms; and at even higher levels may produce outflow-dominant storms.

Suggested values are:
600-900 j/kg: Threshold for damaging wind
>1100 j/kg: Damaging wind

5. Lifted Index (LI)

The Lifted Index was widely favored in the 1980s before integrated stability measures became widespread. It lifts a parcel from the surface to 500 mb and compares the parcel temperature to the environmental temperature. The parcel should be lifted based on the forecast afternoon temperature and dewpoint. The historical origin of the Lifted Index is attributed to Joseph Galway in 1956.

$$LI = T_{ENVIR(500)} - T_{PARCEL(500)}$$

where $T_{PARCEL(500)}$ is that of a surface-based parcel (SBLI), from a mixed layer (MLLI), preferably the lowest 100 mb, or from a pressure level that yields the lowest, or most unstable, lifted index (MULI)

Typical values are:

>2 - No significant activity
2 to 0: Showers/thunderstorms possible with other source of lift
0 to -2: Thunderstorms possible
-2 to -4: Thunderstorms probable, only a few severe
<-4: Severe thunderstorms possible

Also:

0 to -2: Weak instability
-3 to -5: Moderate instability
-6 to -9: Strong instability
-10 or less: Extreme instability

6. Modified Lifted Index (MLI)

The Modified Lifted Index is the same as the Lifted Index except that the parcel is lifted from the highest wet bulb temperature in the lowest 300 mb of the atmosphere. The index was developed by Charles Doswell to better forecast thunderstorms on the High Plains. It was presented as a possible local enhancement to the lifted index, but was adopted for use by the NWS and military forecasters.

$$MLI = T_{ENVIR(500)} - T_{PARCEL(500)}$$

parcel lifted from maximum wet bulb temperature in lowest 300 mb of atmosphere

Typical values are:

Positive: No thunderstorms likely

0 to -2: Showers probable, thunderstorms possible
-3 to -5: Moderate indication of thunderstorms
<-6: Strong indication of severe thunderstorms

Another type of modified lifted index exists which raises a parcel to the -20 deg C isotherm instead of to the 500 mb level. The underlying concept is that a significant thunderstorm should have a cloud temperature of -20 deg C. The historical source of this index is not known.

7. Showalter Stability Index (SSI)

The SSI lifts a parcel from 850 mb to 500 mb. It has the advantage of avoiding the problems inherent with shallow moisture situations, however this comes at the cost of ignoring boundary-layer characteristics. It does not work well in mountainous areas, and cannot be used when the 850 mb level is below ground level. The SSI was developed by Albert Showalter in 1947.

$$SSI = T_{500(ENVIR)} - T_{500(PARCEL)} \quad \text{(parcel lifted from 850 mb)}$$

Typical values are:
>+3: No thunderstorms likely
+3 to +1: Showers probable, thunderstorms possible
0 to –3: Moderate indication of severe thunderstorms
–4 to –6: Strong indication of severe thunderstorms
<-6: Severe thunderstorms likely

8. Vertical Totals Index (VT)

The Vertical Totals Index (Miller 1972) is a measure of the lapse rate from about 5,000 to about 18,000 ft in the atmosphere. It makes no assumptions about parcel temperature. The VT Index was published by Robert Miller in 1967.

$$VTI = T_{850} - T_{500}$$

Typical values are:
25 Storms are unlikely
26 Scattered thunderstorms
30 Scattered thunderstorms, a few severe, isolated tornadoes
32 Scattered to numerous thunderstorms, scattered to a few severe, a few tornadoes
34+ Numerous thunderstorms, scattered severe storms, scattered tornadoes

9. Cross Totals Index (CT)

The Cross Totals Index (Miller 1972) relates the low-level moisture to the mid-level temperature, yielding an indirect estimate of lapse rate and convective instability. It was published by Robert Miller in 1967. The Surface Based Cross Totals Index (SCTI) is an alternative index that uses the surface instead of 850 mb.

$$CTI = T_{850} - T_{500}$$

Typical values are:

<17 - Thunderstorms unlikely
18 to 19 - Isolated to few thunderstorms
20 to 21 - Scattered thunderstorms
22 to 23 - Scattered thunderstorms, isolated severe
24 to 25 - Scattered thunderstorms, few severe, isolated tornadoes
26 to 29 - Scattered to numerous thunderstorms, few to scattered severe, few tornadoes
30+ - Numerous thunderstorms, scattered severe, scattered tornadoes

10. Total Totals Index (TT)

The Total Totals Index (Miller 1972) attempts to integrate the lapse rate information of the Vertical Totals Index with the instability information in the Cross Totals Index. The index is sensitive to steep lapse rates, even if insufficient moisture is present. It was devised by Robert Miller in 1967.

$$TT = VT + CT$$

Typical values are:

<43 - Thunderstorms unlikely
44-45 - Isolated or few thunderstorms
46-47 - Scattered thunderstorms
48-49 - Scattered thunderstorms, isolated severe
50-51 - Scattered heavy thunderstorms, few severe, isolated tornadoes
52-55 - Scattered to numerous heavy thunderstorms, few to scattered severe, few tornadoes
56+ - Numerous heavy thunderstorms, scattered severe, scattered tornadoes

11. K Index (KI)

The K Index (George 1960) is a sum of lapse rate, 850 mb moisture, and humidity at 700 mb. Humidity at 700 mb is a significant contribution, though this is rare with Great Plains storm events or any event that depends on a mid-level cap. The index was published by J. J. George in 1960.

$$KI = T_{850} - T_{500} + T_{d\,850} - D_{700}$$

Typical values are:

<15 - No thunderstorms (0%)
15-20 - Thunderstorms unlikely (<20%)
21-25 - Isolated thunderstorms (20-40%)
26-30 - Widely sct. thunderstorms (40-60%)
30-35 - Numerous thunderstorms (60-80%)
36-40 - Numerous thunderstorms (80-90%)
40+ - Definite thunderstorms (100%)

12. Thompson Index (TI)

The Thompson Index is a combination of Lifted Index and K Index. It attempts to integrate elevated moisture into the index, using the 850 mb dewpoint and 700 mb humidity. Accordingly, it works best in tropical and mountainous locations. The historical origin of the index is not known.

$$TI = K - LI$$

Over Rockies

<20: Thunderstorms unlikely
20-29: Thunderstorms
30-34: Thunderstorms approaching severe
35+: Severe thunderstorms

East of Rockies

<25: Thunderstorms unlikely
25-34: Slight chance of thunderstorms
35-39: Few widely scattered thunderstorms approaching severe
>40: Severe thunderstorms

13. Severe Weather Threat Index (SWEAT)

The SWEAT index uses a complex set of parameters. It was one of the first indices developed specifically to assess tornado potential. Important parameters are 850 mb dewpoints, lapse rates and parcel instability, wind speed at 850 mb and 500 mb, and directional shear betwen 850 mb and 500 mb. The SWEAT index was published in 1972 by Robert Miller. The index uses the Total Totals Index, which must be computed first.

$$SWEAT = 12 \times T_{d\,850} + 20 \times (TT\text{-}49) + 2 \times FF_{850} + FF_{500} + 125 \times (\sin[DD_{500} - DD_{850}] + 0.2)$$

Typical values are:

<300: Non-severe thunderstorms
300-400: Isolated moderate to heavy thunderstorms
400-500: Severe thunderstorms and tornadoes probable
500-800: Severe thunderstorms and tornadoes likely
800+: Possibly no severe weather (sheared convection)

14. Significant Tornado Parameter (STP)

The STP index was developed by the Storm Prediction Center. It is similar to the SWEAT index in that it uses a large number of parameters. The formulation is:

STP = (MLCAPE/1500) * ((2000-MLLCL)/1500) * (ESRH/150) * (ESHEAR/20) * ((200+MLCIN)/150)

where MLCAPE is the lowest 100 mb mixed layer CAPE in j/kg, MLLCL is the lowest 100 mb mixed parcel LCL height in meters, MLCIN is the lowest 100 mb mixed layer CINH in j/kg, ESRH is the effective storm-relative helicity, and ESHEAR is the maximum bulk shear from the most unstable parcel level upward. If the MLLCL is less than 1000 m, the entire MLLCL term is set to 1. If the MLCIN is greater than -50 j/kg, the entire MLCIN term is set to 1. The entire ESHEAR term cannot exceed 1.5, and is set to 0 if ESHEAR is less than 12.5.

Values of STP are as follows:

0.4: Median value associated with nontornadic storms
0.9: Median value associated with weak tornadoes
2.7: Median value associated with significant tornadoes

15. TQ Index (TQ)

The TQ Index is used for assessing the probability of low-topped convection:

$$TQ = (T_{850} + T_{d\,850}) - 1.7(T_{700})$$

Values of more than 12 indicate an unstable lower troposphere with thunderstorms possible outside of stratiform clouds, and val-

Forecasting

ues of more than 17 indicate an unstable lower troposphere with thunderstorms possible when stratiform clouds are present.

16. Delta theta-e

The Delta Theta-e index (Atkins and Wakimoto 1991) is used for assessing the probability of wet microbursts. The formula is:

$$\mathbf{DT} = \theta_{e\,S} - \theta_{e\,U}$$

where $\theta_{e\,S}$ is the equivalent potential temperature at the surface and $\theta_{e\,U}$ is the lowest equivalent potential temperature in the mid-levels of the troposphere.

17. Bulk Richardson Number (BRN)

The Bulk Richardson Number is a ratio between instability and 0-6 km vertical shear. It is a discriminator of storm type, not a predictor. High values indicate unstable and/or weakly sheared environments, while low values indicate weak instability and/or strong shear. It was defined in 1986 by Morris Weisman and Joseph Klemp.

$$BRN = CAPE \,/\, [0.5 \times U^2]$$

where U is the difference in wind speed in meters per second between 0 and 6 km

Typical values are:

<10: Severe weather unlikely
10-45: Associated with supercell development
>50: Weak multicell storms

18. BRN Shear

BRN shear is simply a measure of the vector difference in the winds through the vertical. The greater the BRN shear, the more likely that a thunderstorm downdraft and precipitation cascade will be separated from the updraft.

$$\textbf{BRN Shear} = 0.5\,(U_{AVG})^2$$

where U_{AVG} is the vector difference between the 0 to 6 km AGL winds and the winds in the lowest 0.5 km of the atmosphere

Typical values are:

25-50: Sometimes associated with tornadic storms
50-100: Associated with tornadic storms

19. Energy-Helicity Index (EHI)

EHI is a product of CAPE and 0-6 km shear, thus it is high when either parameter is high. It was developed in 1991 by John Hart and Josh Korotky.

$$EHI = (CAPE \times SRH) / 160{,}000$$

Typical values are:
- 1.0-2.0: Heightened threat of tornadoes
- 2.0-2.4: Tornadoes possible but unlikely strong
- 2.5-2.9: Tornadoes likely
- 3.0-3.9: Strong tornadoes possible
- 4.0+: Violent tornadoes possible

20. Storm Relative Helicity (SRH)

EHI is a product of CAPE and 0-6 km shear, thus it is high when either parameter is high. The SRH is not used widely because it has a high dependence on the correct storm motion vector and it is extremely sensitive to the wind field. The SRH was defined in 1990 by Robert Davies-Jones, Don Burgess, and Mike Foster.

$$SRH = \mathbf{w} \times (\mathbf{v} - \mathbf{c})\, Dz$$

where $\mathbf{w} = \mathbf{k} \times dV/dz$ from 0 to 6 km, $\mathbf{v}$ is the wind vector, and $\mathbf{c}$ is the storm vector

Typical values are:
- 150-299: Weak tornado potential
- 300-449: Moderate tornado potential
- 450+: Strong tornado potential

21. Vorticity Generation Potential (VGP)

The Vorticity Generation Potential index was developed by researcher Erik Rasmussen, and is based on the work of Rasmussen and Wilhelmson (1982). It assesses the possibility for vorticity being tilted into the vertical to create rotating updrafts. The formula is:

$$VGP = \mathrm{sqrt}(CAPE) \times U_{0\text{-}3}$$

where $U_{0\text{-}3}$ is the shear between the 0 and 3 km layer

Suggested interpretations are:
- <0.3: Supercell tornadoes unlikely

Forecasting

>0.6: Supercell tornadoes likely

22. KO Index

This index was developed by Swedish meteorologists and used heavily by the Deutsche Wetterdienst. It compares values of equivalent potential temperature at different levels. It was developed by T. Andersson, M. Andersson, and C. Jacobsson, and S. Nilsson.

$$KO = 0.5 \times (q_{e\,500} + q_{e\,700}) - 0.5 \times (q_{e\,850} + q_{e\,1000})$$

Typical values are:
>6: No thunderstorms
2-6: Thunderstorms possible
<2: Severe thunderstorms possible

23. Boyden Index (BI)

This index, used in Europe, does not factor in moisture. It evaluates thickness and mid-level warmth. It was defined in 1963 by C. J. Boyden.

$$BI = Z_{700} - Z_{1000} - T_{700} - 200$$

where Z is height in decameters (dam)

Typical values are:
>95: Thunder possible

24. Bradbury Index (BRAD)

Also known as the Potential Wet-Bulb Index, this index is used in Europe. It is a measure of the potential instability between 850 and 500 mb. It was defined in 1977 by T. A. M. Bradbury.

$$BRAD = q_{w\,500} - q_{w\,850}$$

Typical values are:
<3: Thunderstorms possible

25. Rackliff Index (RI)

This index, used primarily in Europe during the 1950s, is a simple comparision of the 900 mb wet bulb temperature with the 500 mb temperature. It is believed to have been developed by Peter Rackliff during the 1940s.

$$RI = q_{w\,900} - T_{500}$$

Typical values are:
>30: Thunderstorms possible

26. Jefferson Index (JI)

A European stability index, the Jefferson Index was intended to be an improvement of the Rackliff Index. The change would make it less dependent on temperature. The version in use since the 1960s is a slight modification of G. J. Jefferson's 1963 definition.

$$JI = 1.6 \times q_{w\,850} - T_{500} - 0.5 \times (T_{700} - T_{d\,700}) - 8$$

Typical values are:
>30: Thunderstorms possible

27. S-Index (S)

This European index is a mix of the K Index and Vertical Totals Index. It was designed to be an optimized vertion of the Total Totals Index. The S-Index was developed by the German Military Geophysical Office.

$$S = KI - (T_{500} + A)$$

where A is 0 if the VT is greater than 25, 2 if the VT is between 22 and 25, and 6 if the VT is less than 22.

Typical values are:
<39: No thunderstorms
41-45: Thunderstorms possible
>46: Thunderstorms likely

28. Yonetani Index (YON)

This index was developed by Japanese meteorologist Tsuneharu Yonetani in 1979 to forecast thunderstorms on the Kanto Plain. It provides a measure of conditional instability and low level moisture.

$$YON = 0.966G_L + 2.41(G_U - G_W) + 0.966g - 15$$

The final term is 16.5 instead of 15 if g is less than or equal to 0.57. G is the layer lapse rate, with G_U representing 850-500 mb and G_L representing 900-850 mb, and W is the lapse rate at 850 mb. The term g is the pressure weighted average of the relative humidity in the 900-850 mb layer, ranging from 0-1.

Typical values are:
>0: Thunderstorms likely

Forecasting

29. Potential Instability Index (PII)

This index relates potential instability in the middle atmosphere with thickness. It was proposed by A. J. Van Delden in 2001.

$$PII = (q_{e\,700} - q_{e\,500}) / (Z_{500} - Z_{925})$$

Typical values are:
>0: Thunderstorms likely

30. Deep Convective Index (DCI)

This index is a combination of parcel theta-e at 850 mb and lifted index. This attempts to further improve the lifted index. It was defined by W. R. Barlow in 1993.

$$DCI = T_{850} + T_{d\,850} - LI$$

Typical values are:
10-20: Weak thunderstorms
20-30: Moderate thunderstorms
30+: Strong thunderstorms

C15. FORECASTING

Global circulation

1. El Niño - Southern Oscillation (ENSO)

The El Niño / La Niña phenomenon is a periodic disruption of the ocean-atmosphere circulation in the Pacific Ocean. It is closely tied to a reversal of the tropical atmospheric circulation called the Southern Oscillation, which in turn triggers El Niño.

1.1. **El Niño ("warm" oscillation)**. El Niño refers to a Pacific Ocean circulation in which the tropical easterlies (trade winds) are dampened and cold upwelling is suppressed along the west coasts of the tropical Americas.

1.1.1. FORECAST IMPACT. Though El Niño does not directly cause weather, it does influence the prevailing pattern. The observable effects of El Niño on North America may include:

- ❑ Persistent extended Pacific jet stream
- ❑ General low pressure in the northern Pacific
- ❑ Wet conditions in California and the southern tier U.S.
- ❑ Cool conditions in the U.S. Gulf of Mexico coast region
- ❑ Warm conditions in the northern tier U.S. and west Canada
- ❑ Dry conditions in the Ohio River Valley region
- ❑ Reduced hurricane activity in the tropical North Atlantic
- ❑ Increased hurricane activity in the eastern North Pacific

1.1.2. EL NIÑO YEARS. Historically noteworthy El Niño events included 1790-93, 1828, 1876-78, 1891, 1925-26, 1982-83, and 1997-98. Recent El Niño episodes include late 1986 through early 1988, summer 1991 through spring 1992, spring 1993, late 1994 through early 1995, winter 1997-98, winter 2002-03, and winter 2004-05.

1.2. **La Niña ("cold oscillation")**. La Niña refers to a Pacific Ocean circulation in which the tropical easterlies (trade winds) are strengthened and cold upwelling is promoted along the west coasts of the tropical Americas.

1.2.1. OBSERVABLE EFFECTS. Though La Niña does not directly cause weather, it does influence the prevailing pattern. The observable effects of La Niña on North America may include:

- ❑ A variable Pacific jet stream
- ❑ Tendency for blocking high pressure / ridging in the Pacific
- ❑ Dry conditions in the southern Rockies and Gulf Coast
- ❑ Warm conditions in the southeast quarter of the U.S.
- ❑ Cold conditions in the northern Rockies and Canada Praries
- ❑ Wet conditions in the coastal Pacific Northwest
- ❑ Wet conditions in the Ohio River Valley region
- ❑ Increased hurricane activity in the tropical North Atlantic
- ❑ Reduced hurricane activity in the eastern North Pacific

1.2.2. LA NIÑA YEARS. Recent La Niña events include summer 1988 through spring 1989, autumn 1995 until spring 1996, summer 1998 until spring 2001, and a tendency toward La Niña in winter 2005-06.

2. Teleconnections

2.1. **Arctic Oscillation** (AO). The Arctic Oscillation is a measure of the difference in sea-level pressure or 500 mb height between the northern polar regions and middle latitudes. Starting in the 1980s the AAO has become progressively more positive.

2.1.1. POSITIVE AO. A positive AO is associated with:

- ❑ Warm temperatures in middle latitudes.

2.1.2. NEGATIVE AO. A negative AO is associated with:

- ❑ Cold temperatures in middle latitudes

2.2. **North Atlantic Oscillation** (NAO). The North Atlantic Oscillation is an Atlantic subset of the Arctic Oscillation (see above). The index was predominantly negative during the 1950s and 1960s, and has tended towards positive since 1970.

2.2.1. POSITIVE NAO. A positive NAO index is associated with:

- ❑ Unusually strong subtropical ("Bermuda") high pressure
- ❑ Unusually deep Icelandic low
- ❑ Warm, wet winters in Europe
- ❑ Cold, dry winters in northern Canada and Greenland
- ❑ Mild, wet winters in the eastern United States

2.2.2. NEGATIVE NAO. A negative NAO index is associated with:

- ❑ Unusually weak subtropical ("Bermuda") high pressure
- ❑ Unusually weak Icelandic low

- ❑ Cold, snowy winters in the eastern United States
- ❑ Mild winters in Greenland
- ❑ Moist air in the Mediterranean and cold air in north Europe

2.3. **Pacific - North American Oscillation** (PNA). The Pacific - North American Oscillation index examines the difference in height between points located in the North Pacific and in the Florida Panhandle.

2.3.1. POSITIVE PNA. A positive PNA index is associated with:
- ❑ Above-average heights near Hawaii and western U.S.
- ❑ Below-average heights near Aleutians and southeast U.S.
- ❑ Enhanced East Asian polar jet with exit over west U.S.
- ❑ Warm winters in western Canada and extreme west U.S.
- ❑ Cold winters in the south central and southeast U.S.
- ❑ Dry conditions in the Midwest states

2.3.2. NEGATIVE PNA. A negative PNA index is associated with:
- ❑ Regression of East Asian polar jet into Asia
- ❑ Blocking patterns over northern Pacific and Alaska
- ❑ Split-flow configuration over northern Pacific
- ❑ Cold winters in western Canada and extreme west U.S.
- ❑ Warm winters in the south central and southeast U.S.
- ❑ Wet conditions in the Midwest states

2.4. **Antarctic Oscillation** (AAO); also "Southern Annular Mode" (SAM). The Antarctic Oscillation index examines the difference in zonal mean sea-level pressure between 65 deg S and 45 deg S. The AAO has become slightly more positive since the 1940s.

2.4.1. POSITIVE AAO. A positive AAO index is associated with:
- ❑ Strong polar vortex
- ❑ Low pressure and low heights above Antarctica
- ❑ High heights in the mid-latitudes
- ❑ Intensified westerlies in the Antarctic Ocean
- ❑ Cold conditions in Antarctica; warm over the Peninsula

2.4.2. NEGATIVE AAO. A negative AAO index is associated with:
- ❑ Weak polar vortex
- ❑ High pressure and high heights above Antarctica
- ❑ Low heights in the mid-latitudes
- ❑ Warm conditions in Antarctica

SECTION D

Numerical Prediction

If we knew exactly the laws of nature and the situation of the universe at the initial moment, we could predict exactly the situation of that same universe at a succeeding moment. but even if it were the case that the natural laws had no longer any secret for us, we could still only know the initial situation approximately ... But it may happen that small differences in the initial conditions produce very great ones in the final phenomena.

HENRI POINCARÉ, 1903
French mathematician and physicist

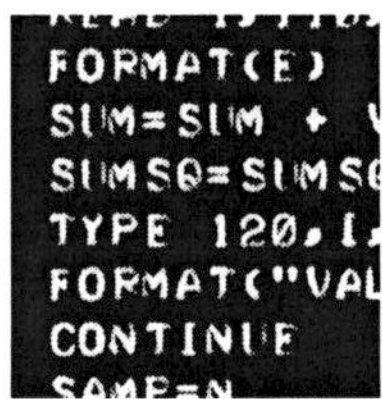

D1. NUMERICAL FORECASTING

Fundamentals

The science of forecasting by models, a process known as *numerical weather prediction* (NWP), can be considered one of the great advances of the 20th century. Since our ability to sample the atmosphere remains very primitive, with one upper-air observation frequently representing regions hundreds or thousands of kilometers in width, all numerical prediction is an approximation. Fortunately there are many different computational methods (models) that can be used to obtain a solution, each of which is comprised of different schemes and algorithms, and each with their own strengths and weaknesses.

1. Forecast production

The process of developing and maintaining an operational numerical weather prediction system with a very high margin of quality is an intensive process and is one reason why only the largest countries have NWP programs. Even private weather services that produce their own model output usually rely on NOAA-derived first-guess and analysis fields.

1.1. **Observation**. Observations are taken by surface stations, radiosonde networks, satellite, profilers, and radar, and are transmitted to central telecommunications facilities.

1.2. **Analysis**. An analysis package (not necessarily part of any specific model) checks the data for errors and interpolates the data onto a grid. Often a *first-guess* field is obtained from previous forecast runs, which helps minimize analysis errors; this is refined with current observations.

1.3. **Initialization**. The initialization process adjusts the analysis and reduces mathematical noise to produce balanced analysis fields. This ensures that the model equations will work successfully.

1.4. **Forecast**. The prediction equations are executed and a forecast is made for a certain number of minutes into the future. This is done repeatedly until a desired forecast time is reached and output is desired.

1.5. **Output**. The forecast grid is produced in a standardized, viewable format. Graphical charts are generated for end users.

2. Model characteristics

2.1. Types of arrays.

2.1.1. GRIDPOINT. Forecast equations are solved at equally-spaced gridpoints. Gridpoint models include the RUC and WRF.

2.1.2. SPECTRAL. Forecast equations are solved in the form of waves representing different amplitudes and wavelengths. Spectral models include the GFS and ECMWF.

2.2. Hydrostatic vs. non-hydrostatic models.

2.2.1. HYDROSTATIC. This type of prediction scheme is commonly used for spectral models and some gridpoint models. This is valid for most synoptic phenomena.

2.2.2. NON-HYDROSTATIC. Commonly used for mesoscale models in which the height of processes are similar to their width (e.g. small scale features). It includes equations for vertical motions not accounted for by hydrostatic models.

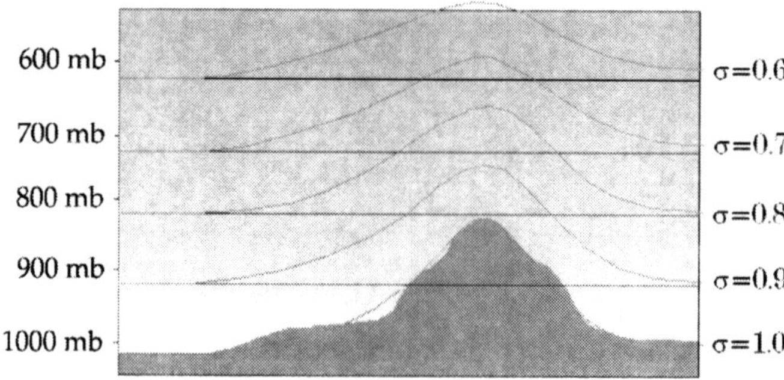

Figure D1-1. The sigma terrain-following coordinate system (curved lines) as viewed against isobars (solid lines). (Adapted from COMET slide)

2.3. **Vertical coordinate systems**

2.3.1. HEIGHT/PRESSURE. The height (km) or pressure (mb/hPa) coordinate system is widely used in manual analysis, but is rarely used in numerical prediction.

2.3.2. THETA (θ) (ISENTROPIC). Surfaces lie along surfaces of equal potential temperature. Since air in adiabatic flow follows these surfaces naturally, the system is suited for numerical prediction. The theta coordinate system has three major problems: (a) isentropes intersect the ground; (b) it poorly handles the boundary layer and diabatic processes such as heating and precipitation, and (c) poor vertical resolution occurs in areas of steep lapse rates.

2.3.3. SIGMA (σ). This coordinate, used by the GFS and MM5, is the most common system in use. It is a *terrain-following coordinate* (Figure D1-1) where the lowest surface $\sigma=1$ coincides with the ground and $\sigma=0$ coincides with the uppermost level of the model. Thus the lowest surfaces follow the Earth's terrain and higher ones are more flat. The greatest problem of sigma levels is in the calculation of pressure gradient along steeply-sloped surfaces. Some issues have been noted in cold-air damming and lee-side cyclogenesis situations.

2.3.4. ETA (η). The eta coordinate system is most widely used in the Eta model. It differs from the sigma system in that terrain is handled as a series of discrete steps, with each step having a width equal to that of the model's horizontal resolution. Each eta surface is flat, even over mountains. The stepped surfaces, however, do not represent terrain very well. A broad area of gently-sloped terrain might be broken into two "flat" areas by the step of an eta surface, while the sigma coordinates would correctly show the gentle slope.

2.3.5. SIGMA-THETA HYBRID. This system, used by the RUC model, uses theta coordinates in the mid- and upper levels of the troposphere and sigma coordinates in the lower troposphere. This scheme retains the advantages of theta coordinates while avoiding terrain intersection and the numerous problems with isentropic coordinates in the boundary layer. However there can be considerable problems blending these coordinate systems at their interface, and processes near the junction may not be represented correctly.

2.4. **Resolution** refers to the spacing between points where prediction calculations are made. It typically takes 3 gridpoints (2 times the resolution) to define the smallest wave and 5 gridpoints (4 times the resolution) to define a feature. Increasing horizontal resolution rapidly escalates the computational work and time needed to solve a prediction.

2.5. **Parameterization** describes physical processes that cannot be individually calculated or measured, or are poorly understood, but which the model seeks to encapsulate in some form. Processes that are parameterized include soil moisture, longwave radiation, insolation, evaporation, convection, cloud processes, and friction. There are an infinite number of ways to parameterize processes in the atmosphere, and it is one of the qualities that makes all models different from one another. Parameterization is also very complicated and it may be difficult to recognize interactions between various parameterization schemes.

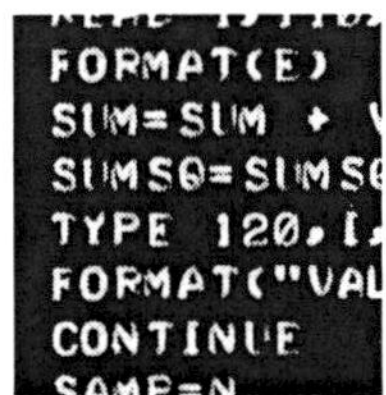

D2. NUMERICAL PREDICTIONS

Model types

This chapter provides an overview of the various meteorological models that a forecaster may encounter. Many of them are operated by the U.S. National Center for Environmental Prediction (NCEP), which as of 2006 continues to be the world's largest distributor of operational forecast data. Model data which is not distributed to the public will not be covered in this section.

1. Operational models, United States

1.1. **WRF (Weather Research & Forecasting)**. Introduced in 2000, the WRF is NOAA's next-generation forecasting system. In the NAM (North American Mesoscale) model slot it replaced the Eta-12 in June 2006. It will replace the RUC in 2007 or 2008, running as the WRF Rapid Refresh. NOAA operates two primary flavors of the WRF: the WRF-MC (Eulerian Mass Core) and WRF-NMM (Non-hydrostatic Mesoscale Model); the latter will be used in most NCEP operation products.
** NCEP NAM (WRF/NMM): Regional domain; Arakawa C-grid (12 or 10 km); 60 hybrid isobaric sigma levels*

1.2. **RUC (Rapid Update Cycle)**. The RUC model is an operational model run by the United States government which gathers a wide variety of data to produce frequent, nationwide short-range forecasts out to 12 hours. It is favored by severe weather forecasters. The RUC will be discontinued in 2007 and replaced with the Rapid Refresh WRF.
** NCEP RUC (June 2005): Regional domain; 13 km gridpoint; 50 isentropic-sigma hybrid levels.*

1.3. **GFS (Global Forecast System) (AVN/MRF)**. The GFS is the preferred model for looking beyond 72 hours, and at conditions around the world. During the 1970s, the National Weather Service relied on the 7LPE (seven-layer primitive equation) model for global prediction). In March 1981 it rolled out a powerful new model known as the GSM, which was renamed MRF and GFS. Beyond 180 hours, the resolution degenerates to 190 waves (about 80 km resolution). As of 2005 it was run out to 384 hours (16 days). The GFS has been cited for deepening mid-tropospheric troughs too much in eastern North America, and not removing elevated instability, which may cause premature development of frontal waves.

* *NCEP GFS (May 05): Global domain; spectral 382-wave (40 km); 50 isentropic-sigma hybrid levels.*

1.4. **NOGAPS (Navy Operational Global Atmospheric Prediction System)**. The U.S. Navy NOGAPS model is arguably the most sophisticated global atmospheric model operated by any military agency. This model was introduced in 1982 as a gridpoint model and was converted to a spectral model in 1988. It is run out to 144 hours.

* *NOGAPS 4.0 (2002): Global domain; 239 waves; 30 sigma levels.*

1.5. **ARPS (Advanced Regional Prediction System)**. ARPS is a mesoscale model developed in the early 1990s by the University of Oklahoma under FORTRAN.

* *ARPS: Mesoscale domain; user-specified Arakawa C-grid with user-specified terrain-following levels*

1.6. **MM5**. The MM5 is a robust mesoscale model developed as a joint project between the University of Pennsylvania and the National Center for Atmospheric Research. It is a limited-area model that can be run in either a hydrostatic or non-hydrostatic configuration. Evaluations have shown the model to be an excellent performer. The concept of the MM5 as an adaptable, customizable model was a great influence on the WRF, which in turn will replace the MM5 among many of its users.

* *MM5: Mesoscale domain; user-specified grid with user-specified terrain-following sigma levels*

1.7. **GEM (Global Environmental Multiscale model)**. The GEM is Canada's primary forecast model. It is run in a Global and a Regional version.

* *GEM Global Forecast System (14 Oct 98): Global domain; 0.9 deg grid; 28 eta levels.*

* *GEM Regional (18 May 04). Regional domain; 15 km grid, 58 sigma levels; 450 sec timestep.*

1.8. **UKMET (United Kingdom Met Office Unified Model)**. The UK Met Office model is a non-hydrostatic gridpoint model covering the entire globe. A mesoscale version of the model covers the United Kingdom. It should be noted that the Met Office restricts all but the basic UKMET forecast fields from public distribution, a policy that is common in European meteorology.
* *UKMET Global: 0.833° long x 0.555° lat (~60 km) Arakawa-C grid; 38 sigma levels.*
* *UKMET Mesoscale: 0.11° (11 km) Arakawa-C grid; 38 levels.*

1.9. **ECMWF IFS (European Centre for Medium-Range Weather Forecasts Integrated Forecast System)**. This European global model was developed in 1985 and during much of the 1990s it was consulted alongside the NOAA MRF to make medium-range forecasts. It should be noted that ECMWF restricts all but the basic forecast fields from public distribution, a policy that is common in European meteorology.
* *ECMWF IFS (Feb 06): Global domain; 799 waves (25 km); 91 levels; 12 min timestep.*

1.10. **MOS (Model Output Statistics)**. MOS, pronounced like the plant "moss", is a broad term covering statistical models which have been developed by NCEP. These models take information from dynamic models above and downscale them to specific locations, yielding tables of temperature, cloud cover, humidity, chance of precipitation, and much more for each location.

1.11. **Ensemble prediction**. Ensemble forecasting is a relatively new technique that draws upon the ability of computers to rapidly execute the full run of a model. Tens, or possibly hundreds or thousands of complete runs are performed, with the parameters, configurations, and starting conditions changing slightly. The output fields can then be compared in many different ways or used to form an *ensemble mean* solution. Typically forecasters will view what is called a "spaghetti plot" to view a few selected isopleths from the entire ensemble package and see the solution and consistency of the entire set.

2. Discontinued models

2.1. **Eta**. The Eta model (named for the Greek letter, not an acronym) was developed by Zaviza Janjic and Fedor Mesinger in Yugoslavia where it became operational in 1978. Its skill at handling the rugged European terrain made it useful in mountainous areas of the United States, especially with East Coast cold air damming situations. However its stepped eta suraces often predicted precipitation too far west, away from mountain peaks, and did not allow enough precipitation on the immediate downwind side of mountain ranges. When implemented by the U.S. in June 1993 it used an 80 km grid with 38 layers, but by 2001 it was running at 12 km resolution with 60 layers. It was retired in June 2006.
* *NCEP Eta (Nov 2001): 12 km grid; 50 isentropic-sigma hybrid levels.*

2.2. **LFM (Limited-area Fine Mesh)**. The LFM model was the first major operational model used by the National Weather Service, introduced in 1971. It was used for nearly all short-term forecast work until the introduction of the NGM model in 1985. The LFM was phased out in 1993.
* *NCEP LFM: 130 km grid; 7 sigma levels.*

2.3. **NGM (Nested Grid Model)**. The NGM, introduced by NMC in 1985, was a unique model that contained a 80 km grid nested within a 190 km grid, in turn nested within a 381 km grid. This formed the backbone of a package known as RAFS (Regional Analysis Forecast System). It was used for nearly all short-term weather forecast work until the introduction of the Eta model in 1993. The NGM was retired in 2003.
* *NMC NGM: 40 km grid; 16 sigma levels*

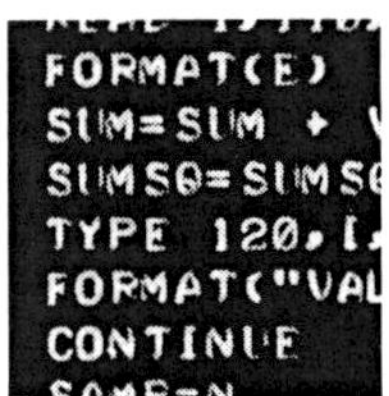

D3. NUMERICAL FORECASTING

Tropical models

1. Operational models

1.1. **GFDL Hurricane Model (GHM).** Developed by the NOAA Geophysical Fluid Dynamics Laboratory (GFDL), the GHM is dynamic baroclinic track prediction model. It uses a 1/12 deg (~9 km) movable grid within a 1/8 deg (~18 km) grid, in turn within a West Hemisphere grid, and with 42 sigma levels (2005). The run is initialized with the storm in the center of the finest grid; the nested grids move with the storm. The run is usually made available about 5 hours after each synoptic observation time.

1.2. **Climatology and Persistence model (CLIPER).** The CLIPER model is a climatology-statistical model. It uses equations which *separately* predict zonal and meridional movement of a tropical cyclone each 12 hours out to 72 hours. Inputs are *current and previous 12 hr position, current and previous 12 hr storm motion, maximum surface winds*, and the *day of the year*. It is often used as a benchmark for dynamical model performance, and was one of the most accurate models through the 1980s. A 5-day version of the model is known as CLIPER5.

1.3. **NHC model 1998 update (NHC98).** This is a dynamical-statistical model which provides an optimum average of three tracks: (a) CLIPER output; (b) equations that separately predict along-track and cross-track motion based on *observed* deep-layer GFS heights; and (c) equations that separately predict along-track and cross-track motion based on *forecast* deep-layer GFS heights. The NHC98 model's predecessors were the NHC90, NHC83, NHC73, NHC72, and NHC67 (the latter two being statistical models).

1.3. **Beta and Advection Model (BAM).** The BAM is a dynamic baroclinic model that uses characteristics of the wind field to steer the predicted hurricane track. The first-guess field is usually run with the previous 12-hour cycle model data, making BAM an early-availability model. There are three versions of the BAM: shallow

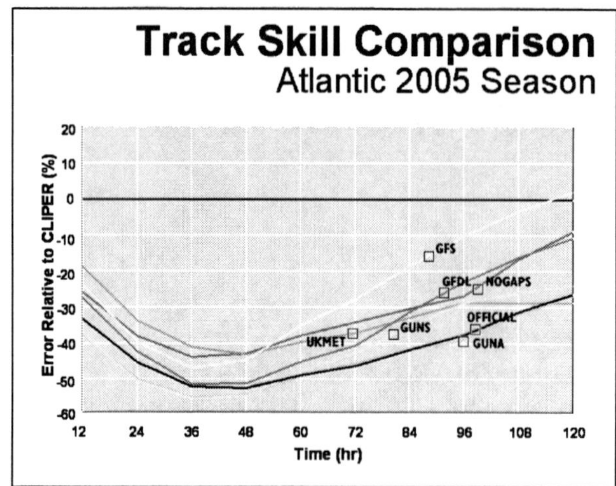

Figure D3-1. Hurricane model performance in 2005. Higher skill is toward the bottom of the page. The GUNA/GUNS ensemble and official forecast were best at all forecast times. Reflecting a consistent pattern in previous years, the GFDL was the best individual model out to 72 hours. Beyond that, the GFS typically showed better skill. In 2005 it did poorly and it was the UKMET which provided the best tracks past 72 hours.

(850-700 mb), medium (850-400 mb), and deep (850-200 mb); as of 2006 it was run 4 times per day. The shallow version is recommended for weak hurricanes, with deeper versions recommended for stronger hurricanes. The BAM was introduced in 1989.

1.4. **NHC Nested Barotropic Hurricane Model (LBAR)**. The LBAR model (an implementation of Vic Ooyama's VICBAR model) is one of the few simple barotropic models in use by NHC. It uses a synoptic domain with initialized fields from the GFS model, a nested regional domain analyzed by LBAR from surface and upper fields, and within that a nested vortex domain which models the hurricane's circulation. The first-guess field is usually run with the previous 12-hour cycle model data, making LBAR an early-availability model. Since it executes very quickly, it often provides the first available forecasts for the storm. It has been noted that LBAR works best early in the hurricane season and on storms that stay on a westward course.

1.5. **Global Forecast System (GFS)**. The GFS is NOAA's primary global forecast model. It is a baroclinic dynamic model with resolution as described in the previous section, and as of July 2006 began using hybrid sigma-pressure coordinates. Most notably it is a spectral model, which uses mathematical waves instead of gridpoints. Starting in 2000, the GFS used a scheme to disassemble the initialized storm vortex and shift it to the official reported positions whenever the storm is over open waters; this replaced the earlier method of introducing the storm directly into the model.

1.6. **Naval Operational Global Atmospheric Prediction System (NOGAPS).** This model is primarily a U.S. Navy global forecast model. However storms are inserted into the model using a set of synthetic soundings based on official estimates of storm location and intensity.

1.7. **GFDL/GFS UKMET NOGAPS Ensemble (CONU).** Created by James Goerss of the Naval Research Laboratory, CONU (pronounced con-you) is an ensemble model that combines the GFDL, the UK Met Office model (which includes hurricane vortex parameterization), the NOGAPS, and the GFS model. Earlier instances were known as GUNS and GUNA. This ensemble has been shown to be slightly more accurate than any individual model; on the other hand it is only more accurate *on average* and may not be appropriate if there is strong divergence between the models.

2. Intensity and storm surge models

2.1. **Statistical Hurricane Intensity Forecast (SHIFOR)** is a statistical model that uses *date, storm intensity, intensity change, position, and storm motion* to find analogs with similar storms dating back to 1967. The SHIFOR model was developed in the 1970s.

2.2. **Statistical Hurricane Intensity Prediction Scheme (SHIPS)** is a statistical-dynamic model that examines the predictors in SHIFOR plus *sea surface temperature*, *vertical shear*, *moist stability*, *200 mb temperature, 850 mb vorticity,* and numerous other parameters in the storm environment to assess its potential for development. It has been in use since 1994. In 2000 a new version of the model, called **Decay SHIP (DSHP)**, was introduced. The DSHP is identical to the SHIPS model except, if the cyclone is forecast to cross land, the intensity is reduced accordingly.

2.3. **Rapid Intensification (RI)** uses output from SHIPS to assess the probability of rapid intensification (24 h increase in maximum wind of 35 mph or greater).

2.4. **Sea Lake and Overland Surges from Hurricanes model (SLOSH)** uses a predefined grid for a coastal area to forecast the flooding caused by an approaching hurricane. It heavily uses bathymetric and topographic data and fluid equations within the water. Small-scale features such as cuts, channels, and levees are parameterized in the model. *The model does not account for astronomical tides, rainfall, river flow, and wind-driven waves.* The model is most often used for planning purposes to define flood-prone areas, but may be run using track and intensity data for an existing storm.

3. Discontinued models

3.1. **Hurricane Analog technique (HURRAN).** The HURRAN is a simple analog model that predicts the storm based on historical tracks. It was developed in 1969, but due to its marginal skill and advances of dynamic models it was discontinued.

D4. NUMERICAL FORECASTING

Pressure methods

1. Weather station sea-level pressure (surface analysis)

Sea-level pressure cannot be directly observed unless the station is at sea-level. In the United States, ASOS sites compute sea-level pressure by taking the average of the current temperature and that 12 hours ago, then extrapolating that to the surface using a standard lapse rate of 6.5 °C/km.

❑ Unusually cold or hot temperatures, a high station elevation, or unusual lapse rates (severe cold or a very unstable air mass) can produce significantly different values than other sea-level pressure estimation methods.

❑ Since the temperature 12 hours ago is factored in, unusual warming or cooling over that period will bias the result

2. NGM/AVN sea-level pressure

The NGM and AVN use the Shuell method of calculating sea-level pressure. It uses the standard lapse rate of 6.5 °C / km. However, the result differs from that produced by ASOS, for several reasons. First, no time-averaging of surface temperature is used. Second, a layer above the model surface is used to determine surface temperature. Finally, restrictions are imposed when the temperature at ground level or 1000 mb is 17.5 °C or greater.

❑ Pressure patterns may be unrepresentative in very warm air masses.

❑ The sea-level pressure fields will not match well to ASOS readings where significant temperature changes are taking place, such as along fronts.

❑ Since the NGM and AVN use a coarse grid, sea-level pressure patterns may differ from that observed due to disparities in surface temperature height.

3. RUC sea-level pressure

The 700 mb temperature, not station temperatures, are used to estimate an "effective" surface temperature. The model extrapolates this 700 mb temperature to the station elevation using the standard atmospheric lapse rate. Although it seems crude, studies have shown that it preserves geostrophic continuity better than other methods. However this may cause problems:

- ❑ The forecast sea-level pressure will not respond to diurnal variations as strongly as expected.
- ❑ Unusual lapse rates (such as inversions, very cold air masses, and very unstable environments) may cause unrepresentative sea-level pressure patterns.

4. ETA sea-level pressure

This algorithm determines the virtual temperature profile beneath a given station down to sea-level by extrapolating *horizontally* from points with lower elevations!

5. LAMP sea-level pressure

The National Weather Service LAMP (Local AWIPS Model Output Statistics Program) sea-level pressure analysis uses actual ASOS sea-level pressure readings, so the output is the same.

6. LAPS sea-level pressure

The National Weather Service LAPS (Local Area Prediction System) tool interpolates ASOS observations to a 10 km grid.

7. MSAS sea-level pressure

The National Weather Service MAPS system uses the MSAS (MAPS Surface Analysis System).

SECTION E

Decoding

E1. DECODING

METAR

WMO FM 15-IX AVIATION SURFACE

METAR stands for Meteorological Airport Report, and is the worldwide standard for transmitting weather reports from airfields. It is the backbone of weather reports in the United States, Europe, and the Pacific Rim.

Familiarity with METAR format is important for a forecaster to be able to pick up on minor trends that might occur at a weather station. The format is:

CCCC ddhhmmZ (AUTO) *dddff VV ww CCCHHH tt/dd P* (RMK)

■ **Station location (*CCCC*)**. This represents the four-letter ICAO identifier where the weather was observed.

■ **Observation time (*ddhhmmZ*)**. The UTC time the observation was taken: calendar day (*dd*); hour (*hh*); and minute (*mm*). The "Z" ending is a reminder that the time zone is Zulu (UTC) time.

■ **Auto flag (AUTO)**. If this flag is present, it indicates the observation was taken by a machine.

■ **Wind (*dddff*)**. The wind direction in degrees relative to true north (*ddd*) and speed (*ff*). If winds are gusting, the group takes the form *dddff*G*gg*, where gg is the gust speed. The group is always appended with units: KT (knots), MPS (meters per second) or KMH (km/h). If the wind direction will be variable, *ddd* is encoded as VRB. It is also permissible to encode the group as *ddd*V*ddd* to indicate a range of wind directions exceeding 60 degrees.

■ **Prevailing visibility (*VV*)**. A number that may be whole or a fraction. Always ends with SM (statute miles) or nothing (meters).

■ **Weather (*ww*)**. A two-letter standard abbreviation (see chart at right) for any weather that will occur, with appropriate modifiers. The term CAVOK may be used if all clouds are above 5000 ft , visibility is above 10 km, and no significant precipitation is occurring (United States does not use CAVOK).

■ **Cloud condition (CCCHHH)**. Assigned for each cloud layer and may repeat. Consists of cloud cover (*CCC*) and height in hundreds of feet (*HHH*). Cloud cover may be clear, few (FEW, 1 to 2 eighths coverage), scattered (SCT, 3 to 4 eighths), broken (BKN, 5 to 7 eighths), or overcast (OVC). When the sky is obscured, *CCC* will be encoded as VV for vertical visibility and the *HHH* value will indicate the visibility into the obscuration.

■ **Temperature/dewpoint (*tt/dd*)**. The temperature (*tt*) and dewpoint (*dd*) in whole degrees Celsius. If any value is negative, it is preceded by an "M".

■ **Pressure (*P*)**. If this value starts with "A", it indicates altimeter setting with the value in hundreds of inches. If the value starts with "Q" it indicates sea-level pressure with the value in whole millibars.

■ **Remarks (RMK)**. If the word "RMK" appears, it indicates that supplementary information follows. Here are some of the more common remarks:

- A01 or A02. Automated station
- SLP*ppp*. Sea-level pressure, where *ppp* is the tens, units, and tenths value in millibars.
- T*atttbddd*. Exact temperature (*tttt*) and dewpoint (*dddd*) in tens of degrees Celsius. The elements *a* and *b* are sign flags: when it is "1" the value that follows it is negative.
- 1*xxxx*. Six-hour max temperature (*xxxx*) in tens of degrees Celsius.
- 2*nnnn*. Six-hour minimum temperature (*nnnn*) in tens of degrees Celsius.
- 4/*sss*. Snow depth (*sss*)in whole inches.
- 4*axxxbnnn*. Twenty-four hour maximum (*xxx*) and minimum (*nnn*) temperature in tens of degrees Celsius. The elements *a* and *b* are sign flags: when it is "1" the value that follows it is negative.
- 5*tppp*. Pressure tendency (*t*) with 0-3 risen, 4 steady, and 5-8 fallen, and change (*ppp*) in tens of millibars.
- 6*pppp*. Six-hour precipitation (*pppp*) in hundreds of inches.
- 7*pppp*. Twenty-four hour precipitation (*pppp*) in hundreds of inches.
- 8/*mh*. Cloud type codes.
- PCPN *pppp* or P *pppp*. One-hour precipitation (*pppp*) in hundreds of inches.

Examples

KBUF 090554Z 06008G21KT 1 1/2SM -SN BR SCT005 BKN021 OVC037 M04/M06 A3003
RMK AO2 PRESFR SLP180 SNINCR 1/2 P0005 60007 4/002 T10441056 11039 21050
58065=

Buffalo, New York on the 9th at 0554 UTC. Winds from 060 true at 8 kt gusting to 21 kt. Visibility 1½ miles in light snow and fog. Scattered layer at 500 ft AGL, broken layer at 2100 ft AGL, overcast layer at 3700 AGL. Temperature minus 4°C, dewpoint minus 6°C. Altimeter setting 30.03 inches. Remarks, automated station. Pressure falling rapidly. Sea level pressure 1018.0 mb. Snow depth increased by ½" in past hour. 1 hr precip 0.05". 6 hr precip 0.07". Snow on ground 2 inches. Maximum 24 hour temperature -4.4°C, minimum -5.6°C. Exact temperature -3.9°C, exact dewpoint -5.0°C, pressure rising then falling, having fallen 6.5 mb.

Figure E1-1. Significant weather, w'w', WMO Code Table 4678. Developed for METAR and TAF use.

Intensity	Description	Precipitation	Obscuration	Other
- Light	**MI** Shallow	**DZ** Drizzle	**FG** Fog (vsby <= 5/8 mile)	**PO** dust/ sand whirls
(no sign) Moderate	**PR** Partial covering of sky	**RA** Rain	**BR** Mist (vsby >= 5/8 mile)	**SQ** Squalls
+ Heavy	**BC** Patches	**SN** Snow	**FU** Smoke	**FC** Funnel cloud, tornado, waterspout
VC Vicinity	**DR** Low drifting	**SG** Snow grains	**VA** Volcanic ash	**SS** Sandstorm
	BL Blowing	**IC** Ice crystals	**DU** Dust	**DS** Dust storm
	SH Showers	**PL** Ice pellets	**SA** Sand	
	TS Thunderstorm	**GR** Hail (>5 mm or >0.2 in)	**HZ** Haze	
	FZ Freezing	**GS** Small hail (<5 mm or <0.2 in)	**PY** Spray	
		UP Unknown		

E2. DECODING

TAF

WMO FM-51 TERMINAL FORECAST

The Terminal Aerodrome Forecast (TAF) is the worldwide standard for encoding standardized forecasts for any airport. It is based on the METAR observation format. For many decades the United States used domestic FT (terminal forecast) style, an extension of their SAO airways observation format. Both the FT and SAO formats were discontinued after a 1993-1995 transition period, and are now historical relics.

A TAF forecast can be particularly useful to meteorologists to ascertain what is expected at another location. The general format is as follows:

CCCC ddhhmm DDHHEE dddff VV ww CCCHHH

■ **Station location (*CCCC*)**. This represents the four-letter ICAO identifier where the forecasted weather will occur.

■ **Issuance time (*ddhhmm*)**. The UTC time the forecast was issued. The calendar day (*dd*); hour (*hh*); and minute (*mm*). A "Z" may be suffixed to the end as a reminder it is Zulu (UTC) time.

■ **Forecast period (*DDHHEE*)**. The UTC time of the forecast period, with the optional starting day (*DD*) and hour (*HH*), and the ending hour (*EE*) (usually on the next day).

■ **Wind (*dddff*)**. The wind direction in degrees relative to true north (*ddd*) and speed (*ff*). If winds are gusting, the group takes the form *dddff*G*gg*, where gg is the gust speed. The group is always appended with units: KT (knots), MPS (meters per second) or KMH (km/h). If the wind direction will be variable, *ddd* is encoded as VRB. It is also permissible to encode the group as *ddd*V*ddd* to indicate a range of wind directions exceeding 60 degrees.

■ **Prevailing visibility (*VV*)**. A number that may be whole or a fraction. Always ends with SM (statute miles) or nothing (meters).

■ **Weather (*ww*)**. A two-letter standard abbreviation for any weather that will occur, with appropriate modifiers. The term CAVOK may be used if all clouds are above 5000 ft , visibility is above 10 km, and no significant precipitation is occurring; the United States does not use CAVOK.

■ **Cloud code group (*CCCHHH*)**. Assigned for each cloud layer and may repeat. Consists of cloud cover (*CCC*) and height in hundreds of feet (*HHH*). Cloud cover may be clear, few (FEW, 1 to 2 eighths coverage), scattered (SCT, 3 to 4 eighths), broken (BKN, 5 to 7 eighths), or overcast (OVC). When the sky is obscured, *CCC* will be encoded as VV for vertical visibility and the *HHH* value will indicate the visibility into the obscuration.

■ **Wind shear group (WS*hhh*/*dddff*)**. Sometimes, particularly in the United States, a low-level wind shear alert will be encoded. The value *hhh* specifies the maximum height above the surface in hundreds of feet, and *ddd* and *ff* specify the wind direction and speed above that height.

■ **Transition identifier**. These introduce new groups of weather conditions that will occur.

- FM *hhmm* indicates a significant change will take place at hour *hh* and minute *mm*.
- TEMPO *hhee* indicates a temporary condition lasting a total of less than half the time period will occur between hour *hh* and hour *ee*.
- BECMG *hhee* indicates a transition period that will begin at hour *hh* and end at hour *ee*, and after this time the new condition will become predominant.
- PROB*pp hhee* indicates a temporary condition with a probability value. The probability in percent is *pp*, and the duration of the expected weather ranges from hour *hh* to hour *ee*. This is used primarily in the United States. Only 30 or 40 is used; if there is a higher probability, then TEMPO is used.

■ **Other groups**. The U.S. military tends to use two groups:

- QNH*pppp*INS, where *pppp* is the lowest expected altimeter setting in hundreds of inches. This is used by U.S. military stations.
- T*tt*/*hh*Z is a maximum and minimum temperature group for the forecast period (two groups are used), where *tt* is the temperature and *hh* is the hour of occurrence. The group may also appear as TN*tt*/*hh*Z TX*tt*/*hh*Z.

Examples

```
BIRK 240412Z 240606 22005KT 9999 SCT020 BKN040 BECMG 1518
          33008KT PROB40 TEMPO 1521 -SHRA BKN015 BECMG 2124
          33015KT 9000 -RA BKN015 OVC030
```

Reykjavik, Iceland, issued on the 24th at 0412 UTC. Forecast period is valid from the 24th at 0600 UTC to 0600 UTC the next day. Winds will be 220 degrees true at 5 kt with unrestricted visibility, scattered layer at 2000 ft AGL and broken at 4000 ft AGL. By 1800 UTC, winds will be 330 degrees true at 8 kt, with a 40% chance of light rain showers and 1500 ft AGL broken clouds between 1500 and 2100 UTC. By 000 UTC winds will be 330 at 15 kt, with visibility 9000 meters in light rain, a broken cloud layer at 1500 ft AGL, and overcast layer at 3000 ft AGL.

Figure E2-1. The TAF code form is specially designed as a universal format to serve the needs of worldwide aviation, transcending most language barriers. *(Alex DeClerk)*

E3. DECODING

SYNOP

WMO FM-12 SYNOPTIC SURFACE

The SYNOP format is given in WMO Pub. 306, "Manual on Codes" in section FM-12. It specifies the format that must be used worldwide for all non-aviation weather observations (aviation reports are coded under METAR format).

The format is as follows:

AAXX *hhmmw*
CCCCC rxhvv NDDFF 1sttt 2SDDD 3pppp 4PPPP 5appp 6RRRT
7wwpp 8NLMH 9hhmm

If any particular group is missing, it is assumed the category is not applicable. For example, if the sky is clear, the 8*NLMH* cloud code group will be omitted. Solidii ("/") usually indicate missing data.

■ **Header (AAXX *hhmmw*)**. The header usually appears at the top of a group of reports collected by a single office. It always starts with "AAXX" ("BBXX" indicates a ship report with more elaborate coding standards that are beyond the scope of this book). The time follows in hours (*hh*) and minutes (*mm*). If the wind indicator (*w*) is 0 or 1, winds are in m/s; if 2 or 3 winds are in knots; an even value is estimated and odd is measured.

■ **Station location (*CCCCC*)**. This represents the five-digit WMO identifier where the weather was observed.

■ **Miscellaneous and visibility group (*rxhvv*)**. The precipitation indicator (*r*) tells whether a supplementary block is used. Station type (*x*) is 1-3 if manned, 4-7 if automated. Lowest cloud height (*h*) is a coded value. Visibility (*vv*) is also a coded value; when below 50 it generally indicates the visibility in tens of kilometers.

■ **Wind group (*NDDFF*)**. This group begins with the total cloud cover (*N*) in eighths; if it is 9 the sky is obscured and if a solidus it is not known. The wind direction in tens of degrees relative to true north (*dd*) and speed (*ff*) are given. The units are given in the header (see above).

■ **Temperature group (*1sttt*)**. Exact temperature (*ttt*) in tens of degrees Celsius. If the sign value (*s*) is "1", the temperature value is negative.

■ **Dewpoint group (*2SDDD*)**. Exact dewpoint temperature (*DDD*) in tens of degrees Celsius. If the sign value (*S*) is "1", the dewpoint value is negative.

■ **Station pressure (3*pppp*)**. The station pressure (*pppp*) is in tens of millibars.

■ **Sea-level pressure (4*PPPP*)**. The sea-level pressure (*pppp*) is in tens of millibars. If the first digit is 1, 2, 5, 7, or 8 this indicates that this is not a sea-level pressure and is instead a geopotential height value.

■ **Pressure tendency (5*appp*)**. Pressure tendency (*a*) with 0-3 risen, 4 steady, and 5-8 fallen, and change (*ppp*) in tens of millibars.

■ **Precipitation (6*RRRT*)**. Liquid precipitation amount (*RRR*) in whole mm. "990" is a trace, and with higher values the last digit is the value in tenths of a mm. The duration of the period (*T*) is "4" if 24 hours, "2" if 12 hours, and "1" if 6 hours.

■ **Weather (7*wwpp*)**. The current weather (*ww*) and past weather (*pp*) is expressed as a two-digit code (see code table). In general the higher the number the more significant the phenomenon.

■ **Cloud group (8*NLMH*)**. The amount of low or middle clouds (*N*) is in eighths. The code for any low (*L*), middle (*M*), or high (*H*) clouds is given.

■ **Time group (9*hhmm*)**. The observation time is given in hours (*hh*) and minutes (*mm*) UTC.

Examples

SMCI07 BABJ 241800 RRA
AAXX 24181
51716 31958 23605 10244 20105 38809 49994 52028 70600 80002
333 00556 10344=

Bachu, China (51716). Manned station, lowest cloud height is coded value 9, visibility is coded value 58. Cloud cover is 2 oktas; winds 360 degrees true at 5 kt. Temperature 24.4°C. Dewpoint 10.5°C. Station pressure 880.9 mb. Sea-level pressure 999.4 mb. Pressure tendency is rising, by 2.8 mb. Weather is dust. Cloud type is dense cirrus.

Figure E3-1. Present weather (ww), WMO Code Table 4677.

Code	Present weather	Code	Present weather
00	Cloud development not observed or observable	50	Drizzle, not freezing, intermittent, slight
01	Clouds dissolving or becoming less developed	51	Drizzle, not freezing, continuous, slight
02	State of sky on the whole unchanged	52	Drizzle, not freezing, intermittent, moderate
03	Clouds generally forming or developing	53	Drizzle, not freezing, continuous, moderate
04	Visibility reduced by smoke haze	54	Drizzle, not freezing, intermittent, heavy
05	Haze	55	Drizzle, not freezing, continuous, heavy
06	Widespread dust in suspension in the air	56	Drizzle, freezing, slight
07	Dust or sand raised by the wind at or near the station	57	Drizzle, freezing, moderate or heavy (dense)
08	Well developed dust whirl(s) or sand whirl(s)	58	Drizzle and rain, slight
09	Duststorm or sandstorm within sight	59	Drizzle and rain, moderate or heavy
10	Mist	60	Rain, not freezing, intermittent, slight
11	Patches of shallow fog or ice fog	61	Rain, not freezing, continuous, slight
12	More or less continuous shallow fog or ice fog	62	Rain, not freezing, intermittent, moderate
13	Lightning visible, no thunder heard	63	Rain, not freezing, continuous, moderate
14	Precipitation within sight, not reaching the ground	64	Rain, not freezing, intermittent, heavy
15	Precipitation within sight, reaching ground	65	Rain, not freezing, continuous, heavy
16	Precipitation, reaching the ground, not at the station	66	Rain, freezing, slight
17	Thunderstorm, but no precipitation	67	Rain, freezing, moderate or heavy
18	Squalls, at or within sight of the station	68	Rain or drizzle and snow, slight
19	Funnel cloud(s) or tornadoes	69	Rain or drizzle and snow, moderate or heavy
20	Recent drizzle (not freezing) or snow grains }	70	Intermittent fall of snowflakes, slight
21	Recent rain (not freezing)	71	Continuous fall of snowflakes, slight
22	Recent snow	72	Intermittent fall of snowflakes, moderate
23	Recent rain and snow or recent ice pellets	73	Continuous fall of snowflakes, moderate
24	Recent freezing drizzle or recent freezing rain	74	Intermittent fall of snowflakes, heavy
25	Recent shower(s) of rain	75	Continuous fall of snowflakes, heavy
26	Recent shower(s) of snow, or of rain and snow	76	Diamond dust (with or without fog)
27	Recent shower(s) of hail, or of rain and hail	77	Snow grains (with or without fog)
28	Recent fog or ice fog	78	Isolated star-like snow crystals (with or without fog)
29	Recent thunderstorm (with or without precipitation)	79	Ice pellets
30	Slight or moderate duststorm	80	Rain shower(s), slight
31	Moderate duststorm	81	Rain shower(s), moderate or heavy
32	Moderate dust storm	82	Rain shower(s), violent
33	Severe dust storm	83	Shower(s) of rain and snow mixed, slight
34	Severe dust storm	84	Shower(s) of rain and snow mixed, moderate or heavy
35	Severe dust storm	85	Snow shower(s), slight
36	Slight or moderate drifting snow, below eye level	86	Snow shower(s), moderate or heavy
37	Heavy drifting snow, below eye level	87	Shower(s) of snow pellets or small hail, slight
38	Slight or moderate blowing snow, above eye level	88	Shower(s) of snow pellets or small hail, moderate/heavy
39	Heavy blowing snow, above eye level	89	Shower(s) of hail, with or without rain, slight
40	Fog or ice fog at a distance	90	Shower(s) of hail, with or without rain, moderate/heavy
41	Fog or ice fog in patches	91	Slight rain at time of observation
42	Fog or ice fog, sky visible, becoming thinner	92	Moderate or heavy rain at time of observation
43	Fog or ice fog, sky obscured, becoming thinner	93	Slight snow, or rain and snow mixed, or hail
44	Fog or ice fog, sky visible, no appreciable change	94	Moderate or heavy snow, or rain and snow, or hail
45	Fog or ice fog, sky obscured, no appreciable change	95	Thunderstorm, slight or moderate, rain, without hail
46	Fog or ice fog, sky visible, thickening	96	Thunderstorm, slight or moderate, with hail
47	Fog or ice fog, sky obscured, thickening	97	Thunderstorm, heavy, rain, without hail
48	Fog or ice fog, sky visible	98	Thunderstorm combined with duststorm or sandstorm
49	Fog or ice fog, sky obscured	99	Thunderstorm, heavy, with hail

E4. DECODING

TEMP

WMO FM-35 UPPER AIR

Nearly all radiosonde data is transmitted in the TEMP format prescribed by the WMO in Publication 306, Section FM-35. It is broken up into three major blocks, TTAA (significant level), TTBB (mandatory level), and PPBB (winds aloft) data. Other blocks such as TTCC and TTDD pertain to data in the stratosphere and is not generally used by forecasters.

■ **Mandatory level block (TTAA)**. This block shows wind, temperature, and dewpoint at predesignated levels, such as 200 and 500 mb.

The block usually begins with:

TTAA *ddhhi ccccc*

The TTAA is a flag that shows this is the mandatory level block. Then follows the calendar day (*dd*) and hour (*hh*). A value of 50 is added to the day if the wind units are in knots, otherwise the wind units are in m/s. The highest wind data (*i*) is a coded figure which is roughly in hundreds of millibars. Finally the WMO station identifier (*ccccc*) is indicated.

Following this is a series of repeating blocks in the format:

pphhh TTTDD dddff

The level (*pp*) is expressed in tens of millibars, e.g. "85" indicates 850 mb. The exception is "99", which always is the first block and indicates the ground, and "92", which is 925 mb, and "88" and "77" are special use (see below). Following this is the height (*hhh*) in different expressions of meters, except for level "99" (ground) in which *hhh* is the surface pressure in tens, units, and tenths of a millibar. The *hhh* expression is in whole meters from 1000 to 700 mb (it is 1*hhh* meters at 850 mb and 2*hhh* or 3*hhh* meters at 700 mb, whichever brings it closer to 3000 m). From 500 to 400 mb *hhh* is expressed in decameters. From 300 to 100 mb *hhh* is 1*hhh* decameters. Following this is the temperature block *TTTDD*, with temperature (*TTT*) in tens of degrees Celsius and dewpoint depression (*DD*) in units and tenths of degrees Celsius if at or below "50" and in whole degrees Celsius if above "50" (subtract 50 before using). Finally the wind is presented as direction (*ddd*) relative to true north and speed (*ff*). Direction always ends with "0" or "5", and "1" is added to it for each hundred units of wind speed (e.g. a *dddff* of 25604 indicates a wind from 255° at a speed of 104).

Tropopause information is encoded as **88*ppp TTTDD dddff*** which indicates conditions at the tropopause: most importantly its pressure. Maximum winds are encoded as ***77ppp dddff***.

■ **Significant level block (TTBB)**. The significant level block is designed to show temperature and dewpoint only at levels bounded by strong changes. The header and format is much the same as the mandatory level (TTAA) block, except that the repeating data block is in the format:

nnppp TTTDD

where *nn* occurs in a repeating sequence (00 for the surface, followed by 11, 22, 33, 44, 55, 66, 77, 88, 99, 11, 22, etc). The rest of the block is identical to the TTAA block with the omission of wind data, and no tropopause or maximum wind data.

■ **Winds aloft block (PPBB)**. This block contains wind data only, and it is graduated in feet rather than millibars. Again, the header and format are similar to TTAA and TTBB format, except that repeating data is in the format:

9habc aaaAA bbbBB cccCC

The "9" is a marker that makes it easy to pick out the group elements. The rest of the group 9*habc* indicates heights of the block, followed by wind data *aaaAA bbbBB cccCC* (encoded the same way as in the TTAA/TTBB sections) at three levels. The ten-thousands place for height is indicated by *h* and the thousands place for each of the three groups by *a*, *b*, and *c*. The height value *ha*000 ft applies to wind group *aaaAA*, *hb*000 ft applies to wind group *bbbBB*, and *hc*000 ft applies to wind group *cccCC*. Not all three groups need to be encoded; if one or two are omitted, *a*, *b*, or *c* will contain a solidus.

Examples

```
520
USUS50 KWBC 241200 RRC

TTAA 74121 72318 99944 13606 26005 00145 ///// ///// 92814 14421
28513 85525 12657 30015 70126 02256 25526 50577 13563 24026 40743
23777 23056 30948 36173 22089 25072 44569 22101 20218 55167 21602
15401 55770 22556 10657 60771 21021 88183 56767 22083 77262 22105
41606 51515 10164 00004 10194 30015 27018=

647
UMUS41 KRNK 241217
SGLRNK

72318 TTBB  74120 72318 00944 13606 11931 14822 22877 11624
33869 13057 44850 12657 55700 02256 66631 01763 77620 02760
88578 05563 99509 12561 11470 17165 22430 19981 33196 55966
44112 56772 55100 60771 31313 45202 81106 41414 50961=

PPBB  74120 72318 90034 26005 30518 31517 90678 28515 28016
25517 909// 25522 91124 25523 25020 23519 916// 24023 92058
24525 23059 22582 93045 22587 22105 22102 9404/ 21602 22068
95024 22536 23528 21521=
```

TTAA: English units, 24th of the month, 1200 UTC. Roanoke, Virginia (72318). At the surface, station pressure is 944 mb, temperature 13.6°C, dewpoint depression 0.6°, winds 260 true at 5 kt. At 1000 mb, height is 145 m, no temperature or dewpoint. At 925 mb, height is 814 m, temperature is 14.4°C, dewpoint depression 2.1°C, winds 285 at 13 kt. At 850 mb, height is 1525 m, temperature is 12.6°C, dewpoint depression 7°, wind 300 at 15 kt, etc.

TTBB: Roanoke, Virginia (72318). English units, 24th of the month, 1200 UTC. At 944 mb, temperature is 13.6°C, dewpoint depression 0.6°. At 931 mb, temperature is 14.8°C, dewpoint depression is 2.2°C. At 877 mb, temperature is 11.6°C, dewpoint depression is 2.4°C.

PPBB: Roanoke, Virginia (72318). At the surface, winds are 260 at 5 kt. At 3000 ft, winds are 305 deg at 18 kt. At 4000 ft, winds are 315 deg at 17 kt. At 6000 ft, winds are 285 deg at 15 kt. At 7000 ft winds are 280 deg at 16 kt. At 8000 ft winds are 255 deg at 17 kt.

SECTION F

Reference

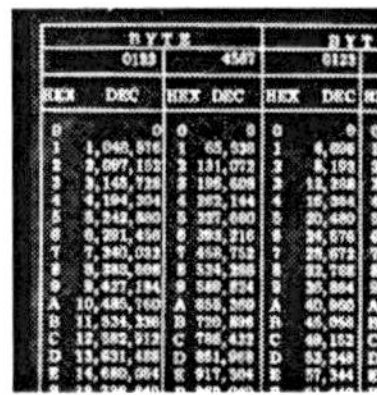

F1. REFERENCE

Constants & variables

A material surface; a part of deformation

a mean radius of the earth; acceleration

B a part of deformation

C surface friction coefficient

c speed of light ($3x10^8$ m s^{-1}); phase speed of a wave

c_p specific heat at a constant pressure

c_v specific heat at a constant volume

E energy of instability

E_{kin} kinetic energy per unit mass

e_s saturation vapor pressure

e_{si} saturation vapor pressure over ice

F friction force per unit mass

F_g geostrophic frontogenesis

f Coriolis parameter ($2\Omega \sin\phi$)

g acceleration of gravity

H relative helicity; height of cold air

h function of pressure; thickness; vertical distance between two surfaces; Planck's constant

K Kelvin; kinematic eddy exchange coefficient

k streamline curvature; Boltzmann's constant

L latent heat of condensation

l coordinate along streamline; wavelength of radar radiation

M Montgomery stream function

m length; map magnification factor

n normal coordinate; constant of the cone; frequency of electromagnetic radiation

P potential vorticity

p air pressure; vertical coordinate

p_0 standard surface pressure; limit of integration in pressure

p_e equilibrium level

Q Q vector

q specific humidity; distance from the equator

q_s saturation specific humidity

R distance from the axis of rotation; specific gas constant for dry air

R_D gas constant of dry air

R_E radius of earth ($6.37x10^6$ m)

r radius of streamline curvature; relative humidity; radius of geographical parallel

Ri Richardson number

R_v specific gas constant for water vapor

S lifted index of stability; image scale of a geographical map

s general space coordinate; slope of front in *p* system; map scale; second

T temperature

Reference

t time coordinate; temperature in °C

T_a ambient temperature

T_d dewpoint

T_l lifting temperature

T_v virtual temperature

u x-component of wind

u_{ag} x-component of ageostrophic wind

u_g x-component of geostrophic wind

u_{gr} x-component of gradient wind

V wind speed

V wind velocity vector

v y-component of wind

$\mathbf{V}_3$ three-dimensional velocity vector

$\mathbf{V}_a$ ageostrophic part of wind

v_{ag} y-component of $\mathbf{V}_3$

$\mathbf{V}_g$ geostrophic wind vector

v_g y-component of geostrophic wind

$\mathbf{V}_{gr}$ gradient wind vector

v_{gr} y-compoennt of gradient wind

V_n normal component of wind

$\mathbf{V}_T$ thermal wind vector

vt tangential component of wind

w vertical component of wind; mixing ratio

w_s saturation mixing ratio

x, y horizontal coordinates

$\mathbf{x}_3$ three-dimensional position vector

x_s,y_s horizontal spherical coordinates

z vertical coordinate (height); height of the isobaric surface; height in geopotential meters

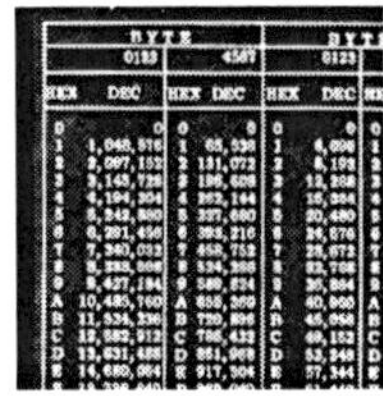

F2. REFERENCE

Greek constants

ALPHABET

Α α	alpha
Β β	beta
Γ γ	gamma
Δ δ	delta
Ε ε	epsilon
Ζ ζ	zeta
Η η	eta
Θ θ	theta
Ι ι	iota
Κ κ	kappa
Λ λ	lambda
Μ μ	mu
Ν ν	nu
Ξ ξ	xi
Ο ο	omicron
Π π	pi
Ρ ρ	rho
Σ σ	sigma
Τ τ	tau
Υ υ	upsilon
Φ φ	phi
Χ χ	chi
Ψ ψ	psi
Ω ω	omega

CONSTANTS AND VARIABLES

α azimuth; general angle; specific volume; planetary albedo

β, β' direction of dilatation axis; direction of air flow

Γ lapse rate of temperature

Γ_a adiabatic lapse rate of temperature

δ polar angle; small vertical distance

ζ vorticity

η absolute vorticity

θ potential temperature

θ_w wet-bulb potential temperature

κ ratio R/c_p

λ longitude

ρ air density

ρ_d density of dry air

ρ_v density of water vapor

σ nondimensional vertical coordinate; Stefan-Boltzmann constant

φ or ϕ geographical latitude

χ velocity potential

ψ stream function; slope of front in z-system; colatitude

Ω angular speed of rotation of the Earth (7.29×10^{-5} rad s^{-1})

ω p-component of wind

Reference

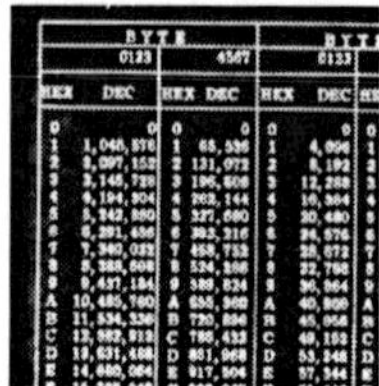

F3. REFERENCE

Time conversion

To convert from one time to another, find the time zone you are converting from. Invert the sign and add this figure to get UTC (for example, if it is 8:45 pm in Moscow, add -3 hours to get 6:45 pm UTC). Then find the time zone you are converting to and add that figure to the UTC time (for example, if U.S. Central Standard Time is desired, add -6 hours to 6:45 pm UTC to get 12:45 pm).

-12:00	Only ships at sea within 7.5° east of 180°
-11:00	American Samoa, Midway Atoll, Niue, Samoa
-10:00	Hawaii-Aleutian Standard Time; Cook Islands, French Polynesia
-9:00	Alaska Standard Time; Gambier Islands
-8:00	Pacific Standard Time; Baja California; Pitcairn Islands
-7:00	Mountain Standard Time
-6:00	Central Standard Time; Central America
-5:00	Eastern Standard Time; Bahamas; Colombia, Cuba; Panama
-4:00	Atlantic Standard Time; Bolivia; Chile; Venezuela
-3:00	Brazil; Argentina; Suriname; Uruguay
-2:00	Fernando de Noronha; South Sandwich Islands
-1:00	Cape Verde; Azores
0:00	UTC; UK; Ireland; Iceland; Liberia; Mali; Mauritania; Morocco
+1:00	Central European Time; Germany; France; Norway; Netherlands
+2:00	Bulgaria; Egypt; Finland; Estonia; Lebanon; Israel; South Africa
+3:00	Moscow; St. Petersburg; Iraq; Kenya; Sudan; Uganda; Saudi Arabia
+3:30	Iran
+4:00	Armenia; Azerbaijan; Georgia; Mauritius; Oman; Seychelles; UAE
+4:30	Afghanistan
+5:00	Maldives; Pakistan; western Kazakhstan; Tajikistan; Uzbekistan
+5:30	India
+5:45	Nepal
+6:00	Bangladesh; Bhutan; eastern Kazakhstan; Kyrgyzstan; Sri Lanka
+6:30	Cocos Islands; Myanmar
+7:00	Cambodia; Laos; Thailand; Vietnam; west Indonesia
+8:00	Australian Western Standard Time; China; Philippines; Singapore
+9:00	Japan; Korea; east Indonesia; East Timor; Palau
+9:30	Australian Central Standard Time
+10:00	Australian Eastern Standard Time; Guam; Papua New Guinea
+10:30	Lord Howe Island
+11:00	New Caledonia; Solomon Islands; Vanuatu
+11:30	Norfolk Island
+12:00	Fiji; Marshall Islands; Nauru; New Zealand; South Pole; Tuvalu
+12:45	Chatham Islands
+13:00	Kiribati's Phoenix Islands; Tonga
+14:00	Kiribati's Line Islands

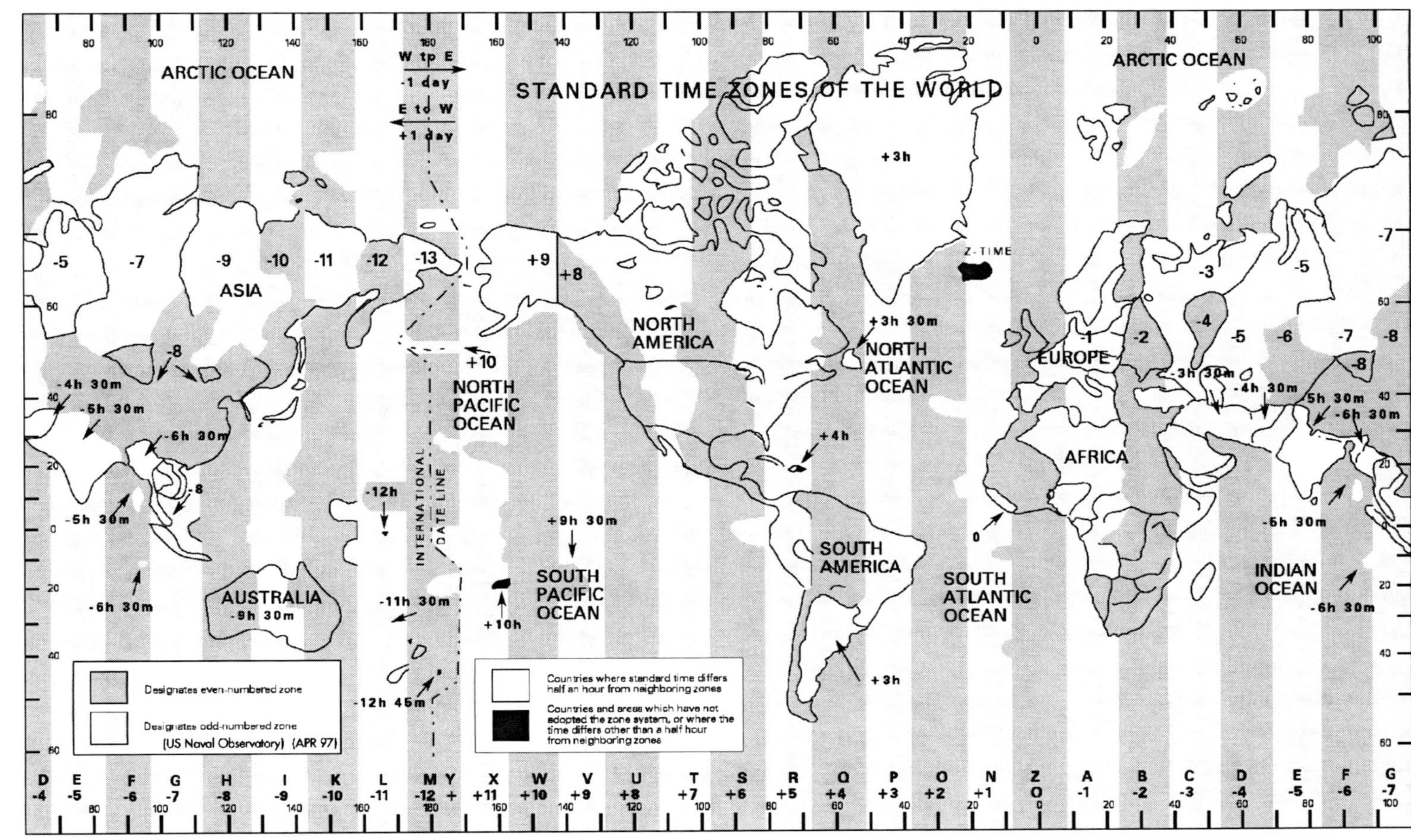
STANDARD TIME ZONES OF THE WORLD
ARCTIC OCEAN
ARCTIC OCEAN
W tp E
-1 day
E to W
+1 day
ASIA
AUSTRALIA
-9h 30m
NORTH AMERICA
SOUTH AMERICA
EUROPE
AFRICA
NORTH PACIFIC OCEAN
SOUTH PACIFIC OCEAN
NORTH ATLANTIC OCEAN
SOUTH ATLANTIC OCEAN
INDIAN OCEAN
INTERNATIONAL DATE LINE
Z-TIME
Designates even-numbered zone
Designates odd-numbered zone
(US Naval Observatory) (APR 97)
Countries where standard time differs half an hour from neighboring zones
Countries and areas which have not adopted the zone system, or where the time differs other than a half hour from neighboring zones
D E F G H I K L M Y X W V U T S R Q P O N Z A B C D E F G
-4 -5 -6 -7 -8 -9 -10 -11 -12 + +11 +10 +9 +8 +7 +6 +5 +4 +3 +2 +1 0 -1 -2 -3 -4 -5 -6 -7

Reference

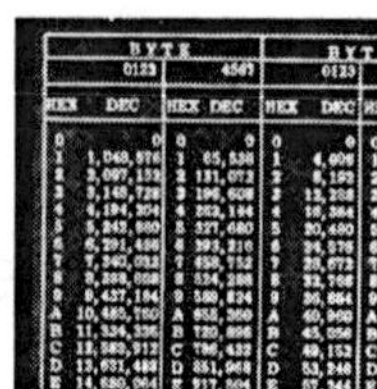

F4. REFERENCE

Unit conversion

1. Wind speed

1.1. Miles per hour (MPH).

= kt × 1.15

= m/s × 2.2356

1.2. Knots (kt).

= MPH × 0.8696

= m/s × 1.944

1.3. Meters per second (m/s).

= kt × 0.5144

= MPH × 0.4473

2. Temperature

2.1. Degrees F (deg F).

= (deg C × 1.8) + 32

= ((K - 273.16) × 1.8) + 32

2.2. Degrees C (deg C).

= (deg F - 32) × 0.555

= K - 273.16

2.3. Kelvin (K).

= deg C + 273.16

= ((deg F - 32) × 0.555) + 273.16

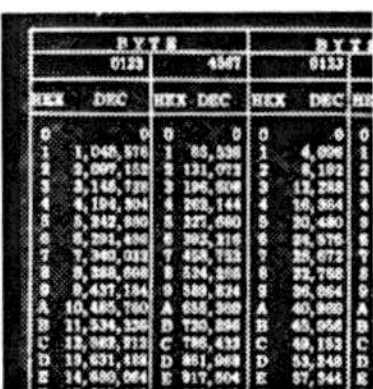

F5. REFERENCE

Pressure conversion

BAROMETRIC READINGS FROM INCHES TO HECTOPASCALS OR MILLIBARS

Inches	0.00	0.01	0.02	0.03	0.04	0.05	0.06	0.07	0.08	0.09
	Hectopascals or Millibars									
28.0-	948.2	948.5	948.9	949.2	949.5	949.9	950.2	950.6	950.9	951.2
28.1-	951.6	951.9	952.3	952.6	952.9	953.3	953.6	953.9	954.3	954.6
28.2-	955.0	955.3	955.6	956.0	956.3	956.7	957.0	957.3	957.7	958.0
28.3-	958.3	958.7	959.0	959.4	959.7	960.0	960.4	960.7	961.1	961.4
28.4-	961.7	962.1	962.4	962.7	963.1	963.4	963.8	964.1	964.4	964.8
28.5-	965.1	965.5	965.8	966.1	966.5	966.8	967.2	967.5	967.8	968.2
28.6-	968.5	968.8	969.2	969.5	969.9	970.2	970.5	970.9	971.2	971.6
28.7-	971.9	972.2	972.6	972.9	973.2	973.6	973.9	974.3	974.6	974.9
28.8-	975.3	975.6	976.0	976.3	976.6	977.0	977.3	977.6	978.0	978.3
28.9-	978.7	979.0	979.3	979.7	980.0	980.4	980.7	981.0	981.4	981.7
29.0-	982.1	982.4	982.7	983.1	983.4	983.7	984.1	984.4	984.8	985.1
29.1-	985.4	985.8	986.1	986.5	986.8	987.1	987.5	987.8	988.1	988.5
29.2-	988.8	989.2	989.5	989.8	990.2	990.5	990.9	991.2	991.5	991.9
29.3-	992.2	992.5	992.9	993.2	993.6	993.9	994.2	994.6	994.9	995.3
29.4-	995.6	995.9	996.3	996.6	997.0	997.3	997.6	998.0	998.3	998.6
29.5-	999.0	999.3	999.7	1000.0	1000.3	1000.7	1001.0	1001.4	1001.7	1002.0
29.6-	1002.4	1002.7	1003.0	1003.4	1003.7	1004.1	1004.4	1004.7	1005.1	1005.4
29.7-	1005.8	1006.1	1006.4	1006.8	1007.1	1007.4	1007.8	1008.1	1005.5	1008.8
29.8-	1009.1	1009.5	1009.8	1010.2	1010.5	1010.8	1011.2	1011.5	1011.9	1012.2
29.9-	1012.5	1012.9	1013.2	1013.5	1013.9	1014.2	1014.6	1014.9	1015.2	1015.6
30.0-	1015.9	1016.3	1016.6	1016.9	1017.3	1017.6	1017.9	1018.3	1018.6	1019.9
30.1-	1019.3	1019.6	1020.0	1020.3	1020.7	1021.0	1021.3	1021.7	1022.0	1022.3
30.2-	1022.7	1023.0	1023.4	1023.7	1024.0	1024.4	1024.7	1025.1	1025.4	1025.7
30.3-	1026.1	1026.4	1026.8	1027.1	1027.4	1027.8	1028.1	1028.4	1028.8	1029.1
30.4-	1029.5	1029.8	1030.1	1030.5	1030.8	1031.2	1031.5	1031.8	1032.2	1032.5
30.5-	1032.8	1033.2	1033.5	1033.9	1034.2	1034.5	1034.9	1035.2	1035.6	1035.9
30.6-	1036.2	1036.6	1036.9	1037.2	1037.6	1037.9	1038.3	1038.6	1038.9	1039.3
30.7-	1039.6	1040.0	1040.3	1040.6	1041.0	1041.3	1041.7	1042.0	1042.3	1042.7
30.8-	1043.0	1043.3	1043.7	1044.0	1044.4	1044.7	1045.0	1045.4	1045.7	1046.1
30.9-	1046.4	1046.7	1047.1	1047.4	1047.7	1048.1	1048.4	1048.8	1049.1	1049.4

Thousandths of an inch

Inches	0.001	0.002	0.003	0.004	0.005	0.006	0.007	0.008	0.009
Hectopascals or Millibars	0.03	0.07	0.10	0.14	0.17	0.20	0.24	0.27	0.30

	BAROMETRIC READINGS FROM MILLIBARS OR HECTOPASCALS TO INCHES									
Hecto-pascals or Millibars	0	1	2	3	4	5	6	7	8	9
	Inches									
940	27.76	27.79	27.82	27.85	27.88	27.91	27.94	27.96	27.99	28.02
950	28.05	28.08	28.11	28.14	28.17	28.20	28.23	28.23	28.29	28.32
960	28.35	28.38	28.41	28.44	28.47	28.50	28.53	28.56	28.59	28.61
970	28.64	28.67	28.70	28.73	28.76	28.79	28.82	28.85	28.88	28.91
980	28.94	28.97	29.00	29.03	29.06	29.09	29.12	29.15	29.18	29.21
990	29.23	29.26	29.29	29.32	29.35	29.38	29.41	29.44	29.47	29.50
1000	29.53	29.56	29.59	29.62	29.65	29.68	29.71	29.74	29.77	29.80
1010	29.83	29.85	29.88	29.91	29.94	29.97	30.00	30.03	30.06	30.09
1020	30.12	30.15	30.18	30.21	30.24	30.27	30.30	30.33	30.36	30.39
1030	30.42	30.45	30.47	30.50	30.53	30.56	30.59	30.62	30.65	30.68
1040	30.71	30.74	30.77	30.80	30.83	30.86	30.89	30.92	30.95	30.98
1050	31.01	31.04	31.07	31.10	31.12	31.15	31.18	31.21	31.24	31.27

	BAROMETRIC READINGS FROM MILLIMETERS TO INCHES									
Milli-meters	0	1	2	3	4	5	6	7	8	9
	Inches									
710	27.97	28.01	28.05	28.09	28.13	28.17	28.21	28.24	28.28	28.32
720	28.36	28.40	28.44	28.48	28.52	28.56	28.60	28.64	28.68	28.72
730	28.76	28.80	28.84	28.88	28.91	28.95	28.99	29.03	29.07	29.11
740	29.15	29.19	29.23	29.27	29.31	29.35	29.39	29.43	29.47	29.51
750	29.55	29.58	29.62	29.66	29.70	29.74	29.78	29.82	29.86	29.90
760	29.94	29.98	30.02	30.06	30.10	30.14	30.18	30.21	30.25	30.29
770	30.33	30.37	30.41	30.45	30.49	30.53	30.57	30.61	30.65	30.69
780	30.73	30.77	30.81	30.85	30.88	30.92	30.96	31.00	31.04	31.08

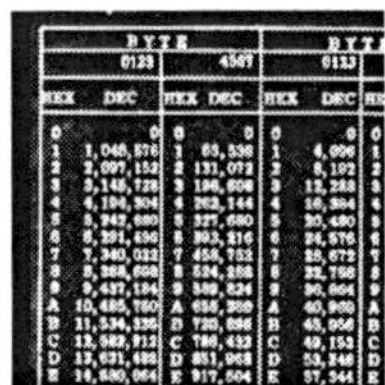

F6. REFERENCE

Heat Index

Heat Index, which is a measure of the sultriness of the air in terms of both temperature and humidity, was introduced in 1979 by R. G. Steadman to replace the antiquated Temperature-Humidity Index (THI). It has come into widespread use in the United States.

It is approximated as follows:

$$HI = -42.379 + 2.04901523T + 10.14333127R - 0.22475541TR - 6.83783\times10^{-3}T^2 - 5.481717\times10^{-2}R^2 + 1.22874\times10^{-3}T^2R + 8.5282\times10^{-4}TR^2 - 1.99\times10^{-6}T^2R^2$$

T = ambient dry bulb temperature

R = relative humidity

TEMPERATURE (deg F)	DEWPOINT (deg F) 58	60	62	64	66	68	70	72	74	76	78	80	82
68	69	70	70	70	70	70							
70	70	71	71	72	72	72							
72	72	72	72	73	73	73	74						
74	73	73	74	75	75	75	76	77					
76	76	76	77	78	78	78	79	80	80				
78	78	78	78	80	80	80	81	82	82				
80	80	80	80	82	82	82	83	84	85	86	87		
82	82	83	84	84	85	86	87	88	89	90	91	93	97
84	84	85	86	86	87	88	89	90	91	93	94	95	99
86	86	87	88	88	89	90	91	91	93	95	96	98	100
88	88	89	90	90	91	91	93	94	95	97	99	100	103
90	90	91	91	91	92	93	95	97	98	100	101	103	106
92	91	92	93	94	95	96	97	99	100	102	104	106	109
94	94	95	96	98	99	100	101	102	104	106	108	110	114
96	96	97	98	100	100	101	103	104	106	109	111	113	117
98	98	99	100	102	103	104	105	107	109	112	114	116	120
100	100	102	103	106	107	108	109	111	113	116	119	122	126
102	102	103	105	108	109	110	112	114	116	119	122		
104	104	105	107	109	111	113	115	117	118	122	124		
106	106	107	109	111	113	115	117	118	121	123			
108	108	109	110	113	115	117	118	120	123				
110	109	110	112	115	117	118	120	122	124				
112	112	113	115	118	119	121	123						
114	114	115	117	119	121	123							
116	116	117	118	121	122								
118	118	119	121	124									

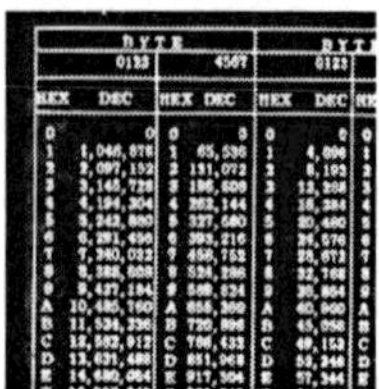

F7. REFERENCE

Fujita damage scale

The Fujita damage scale was developed in 1971 by Theodore Fujita. On February 1, 2007 the United States will begin using the Enhanced F-Scale (EF).

Rating	Wind speed	Original damage description
EF0 F0	**65-85 mph** 40-72 mph	Some damage to chimneys; breaks branches off trees; pushes over shallow-rooted trees; damages sign boards
EF1 F1	**86-109 mph** 73-112 mph	The lower limit is the beginning of hurricane wind speeds; peels surface off roofs; mobile homes pushed off foundations or overturned; moving autos pushed off roads; attached garages may be destroyed
EF2 F2	**110-137 mph** 113-157 mph	Considerable damage. Roofs torn off frame houses; mobile homes demolished; boxcars pushed over; large trees snapped or uprooted; light object missiles generated
EF3 F3	**138-167 mph** 158-206 mph	Roof and some walls torn off well constructed houses; trains overturned; most trees in forest uprooted
EF4 F4	**166-200 mph** 207-260 mph	Well-constructed houses levelled; structures with weak foundations blown off some distance; cars thrown; large missiles generated
EF5 F5	**201+ mph** 261-318 mph	Strong frame houses lifted off foundations and carried considerable distances to disintegrate; automobile-sized missiles fly through air in excess of 100 m; trees debarked; steel reinforced concrete structures badly damaged

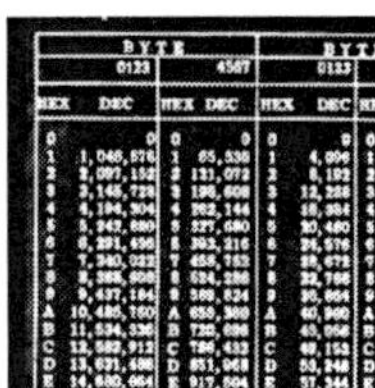

F8. REFERENCE

Hail size

The National Severe Storms Forecast Center strongly encourages measurement, not estimation of hail size. Sizes shown in bold are the most commonly used descriptors.

Diameter		
inches	mm	Descriptor
0.25	6	Pea
0.50	13	**Marble**; Moth ball
0.75	19	Penny; Dime
0.88	22	Nickel
1.00	25	**Quarter**
1.25	32	Half-dollar
1.50	38	Walnut; Ping-pong ball
1.75	44	**Golf ball**
2.00	51	Hen egg
2.50	64	Tennis-ball
2.75	70	**Baseball**
3.00	76	Tea cup
4.00	102	Grapefruit
4.50	114	**Softball**

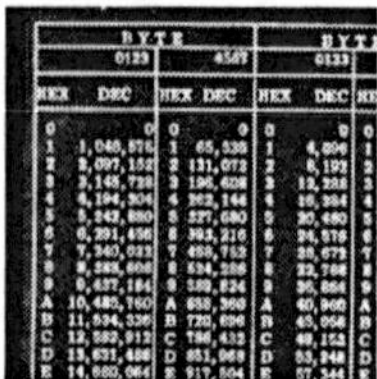

F9. REFERENCE

Qualifiers

Forecasting is not yet advanced enough to predict weather occurrences with any exact accuracy. When communicating information to the public, it has become necessary to use qualifying expressions to indicate probabilities, characteristics, and so forth.

■ **Duration**. A duration qualifier implies a high degree of confidence that the expected weather will occur. The National Weather Service allows the use of "brief periods of", "occasional", and "intermittent".

■ **Intensity**. The National Weather Service allows the use of "very light", "light", "heavy", and "very heavy". No intensity modifier is used for moderate precipitation.

■ **Probability**. In typical use, a probability forecast is strictly the forecaster's degree of confidence that it will occur at a given location. It makes no implied assumption about how extensive the weather will be over a given area. Accepted terminology is as follows:

Chance	Expression of uncertainty	Equivalent convective areal qualifier
0%	(none used)	(none used)
10%	Slight chance	Isolated[2]; few[2]
20%	Slight chance	Widely scattered[2]
30, 40, 50%	Chance	Scattered[2]
60, 70	Likely	Numerous[2]
80, 90, 100%	(none used)[1]	(none used)

[1] The weather condition is stated without qualifiers (e.g. "Rain")
[2] The NWS allows this qualifier when the chance of convective precipitation somewhere in a forecast area is very high.
Source: WSOM C-11, National Weather Service.

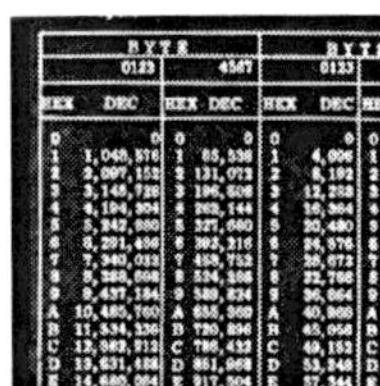

F10. REFERENCE

Radar Intensity

Radar levels, as used during the 1960s, 1970s, and 1980s.

VIP LVL	EQV dBz	INTENSITY	RAINFALL RATE (in/hr) CONVECTIVE	STRATIFORM	CHARACTER
0	-	BASE	None	<0.05"	No precip or drizzle
1	18-29	LIGHT	<0.20"	<0.10"	Light precipitation
2	30-40	MODERATE	0.20-1.10"	0.10-0.50"	Moderate precipitation
3	41-45	HEAVY	1.10-2.20"	0.50-1.00"	Heavy precipitation; thunder
4	46-49	VERY HEAVY	2.20-4.50"	—	Marble-sized hail possible
5	50-56	INTENSE	4.50-7.10"	—	Golf-ball sized hail possible
6	>56	EXTREME	7.10"+	—	Baseball-sized hail possible

Radar levels on the WSR-88D NEXRAD system are as follows:

NIDS LVL	EQV dBz	INTENSITY	RAIN RATE (IN/HR)*	CHARACTER
0	-	BASE	None	No echoes
1	5-10	Marginal	Trace	Collision and coalescence in cloud
2	10-15	Extremely lgt	0.01"	Virga and cloud material
3	15-20	Very light	0.01"	Drizzle or very light precipitation
4	20-25	Light	0.03"	Light showers
5	25-30	Moderate	0.05"	Moderate showers
6	30-35	Heavy	0.11"	Thunderstorms possible
7	35-40	Very Heavy	0.22"	Thunderstorms possible
8	40-45	Very Heavy	0.45"	Small hail is possible
9	45-50	Intense	0.92"	Small hail is possible
10	50-55	Intense	1.89"	Moderate-sized hail is possible
11	55-60	Very Intense	3.89"	Moderate-sized hail is possible
12	60-65	Very intense	8.00"	Large hail is possible
13	65-70	Extreme	15"	Large hail is possible
14	70-75	Extreme	34"	Very large hail is possible
15	>75	Extreme	69"	Very large hail is possible

* Rainfall rate is an approximation given by $[10^{(dBz/10)}/200]^{0.625}$ and is highly dependent on drop size distribution and the presence of hail; values shown are only a guide.

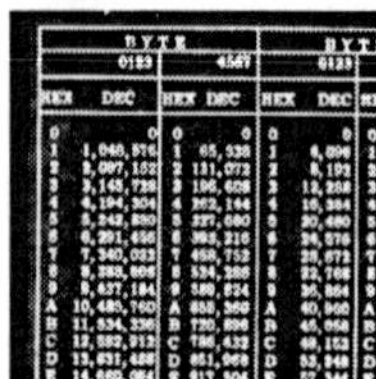

F11. REFERENCE

ICAO

This chart lists the meaning of the first two letters of an ICAO (International Civil Aviation Organization) identifier. (ICAO 2005)

- AG Solomon Islands
- AN Nauru
- AY Papua New Guinea
- BG Greenland
- BI Iceland
- C Canada
- DA Algeria
- DB Benin
- DF Burkina Faso
- DG Ghana
- DI Côte d'Ivoire
- DN Nigeria
- DR Niger
- DT Tunisia
- DX Togolese Rep.
- EB Belgium
- ED Germany (civil)
- EE Estonia
- EF Finland
- EG United Kingdom
- EH Netherlands
- EI Rep. of Ireland
- EK Denmark
- EL Luxembourg
- EN Norway
- EP Poland
- ES Sweden
- ET Germany (mil)
- EV Latvia
- EY Lithuania
- FA South Africa
- FB Botswana
- FC Rep. of Congo
- FD Swaziland
- FE Cen. Afr. Rep.
- FG Equat. Guinea
- FH Ascension Island
- FI Mauritius
- FJ Brit. Ind. Oc. Ter.
- FK Cameroon
- FL Zambia
- FM Comoros, Madagascar, Mayotte, Réunion
- FN Angola
- FO Gabon
- FP São Tomé and Príncipe
- FQ Mozambique
- FS Seychelles
- FT Chad
- FV Zimbabwe
- FW Malawi
- FX Lesotho
- FY Namibia
- FZ Dem. Rep. Congo
- GA Mali
- GB The Gambia
- GC Canary I. (Spain)
- GE Ceuta and Melilla
- GF Sierra Leone
- GG Guinea-Bissau
- GL Liberia
- GM Morocco
- GO Senegal
- GQ Mauritania
- GS Western Sahara
- GU Guinea
- GV Cape Verde
- HA Ethiopia
- HB Burundi
- HC Somalia
- HD Djibouti (also HF)
- HE Egypt
- HF Djibouti (and HD)
- HH Eritrea
- HK Kenya
- HL Libya
- HR Rwanda
- HS Sudan
- HT Tanzania
- HU Uganda
- K United States
- LA Albania
- LB Bulgaria
- LC Cyprus
- LD Croatia
- LE Spain
- LF France, incl. Saint-Pierre
- LG Greece
- LH Hungary
- LI Italy
- LJ Slovenia
- LK Czech Republic
- LL Israel
- LM Malta
- LN Monaco
- LO Austria
- LP Portugal, Azores
- LQ Bosnia, Herzeg.
- LR Romania
- LS Switzerland
- LT Turkey
- LU Moldova
- LV Gaza Strip
- LW Macedonia
- LX Gibraltar
- LY Serbia, Monte.
- LZ Slovakia
- MB Turks and Caicos
- MD Dominican Rep.
- MG Guatemala
- MH Honduras
- MK Jamaica
- MM Mexico
- MN Nicaragua
- MP Panama
- MR Costa Rica
- MS El Salvador
- MT Haiti
- MU Cuba
- MW Cayman Islands
- MY Bahamas
- MZ Belize
- NC Cook Islands
- NF Fiji, Tonga
- NG Kiribati, Tuvalu
- NI Niue
- NL Wallis and Futuna
- NS Samoa
- NT French Polynesia
- NV Vanuatu
- NW New Caledonia
- NZ New Zealand
- OA Afghanistan
- OB Bahrain
- OE Saudi Arabia
- OI Iran
- OJ Jordan and the West Bank
- OK Kuwait
- OL Lebanon
- OM U. Arab Emirates
- OO Oman
- OP Pakistan
- OR Iraq
- OS Syria
- OT Qatar
- OY Yemen
- PA Alaska only
- PB Baker Island
- PC Kiribati
- PF Fort Yukon, Ak.
- PG Guam, N. Mari.
- PH Hawaii only
- PJ Johnston Atoll
- PK Marshall Islands
- PL Kiribati (Line I.)
- PM Midway Island
- PO Oliktok, Alaska
- PP Point Lay, Alaska
- PT Fed. Micronesia
- PW Wake Island
- RC Taiwan
- RJ Japan
- RK South Korea
- RO Japan (Okinawa)
- RP Philippines
- SA Argentina
- SB Brazil
- SC Chile
- SD Brazil
- SE Ecuador
- SF Falkland Islands
- SG Paraguay
- SK Colombia
- SL Bolivia
- SM Suriname
- SN Brazil
- SO French Guiana
- SP Peru
- SS Brazil
- SU Uruguay
- SV Venezuela
- SW Brazil
- SY Guyana
- TA Antigua, Barbuda
- TB Barbados
- TD Dominica
- TF Guadeloupe
- TG Grenada
- TI U.S. Virgin Isl.
- TJ Puerto Rico
- TK Saint Kitts Nevis
- TL Saint Lucia
- TN Nth. Antil., Aruba
- TQ Anguilla
- TR Montserrat
- TT Trinidad, Tobago
- TU Brit. Virgin Isl.
- TV St Vincent, Gren.
- TX Bermuda
- U Russia (except UA, UB, UG, UK, UM and UT)
- UA Kazakh., Kyrgyz.
- UB Azerbaijan
- UG Armenia, Georgia
- UK Ukraine
- UM Belarus
- UT Tajikistan, Turkmenistan, Uzbekistan
- VA India
- VC Sri Lanka
- VD Cambodia
- VE India
- VG Bangladesh
- VH Hong Kong
- VI India
- VL Laos
- VM Macao
- VN Nepal
- VO India
- VQ Bhutan
- VR Maldives
- VT Thailand
- VV Vietnam
- VY Myanmar
- WA Indonesia
- WB Malaysia, Brunei
- WI Indonesia
- WM Malaysia
- WP Timor-Leste
- WQ Indonesia
- WR Indonesia
- WS Singapore
- Y Australia
- Z Peo. Rep. China (except ZK and ZM)
- ZK North Korea
- ZM Mongolia

ICAO Regions

First letter of an ICAO identifier

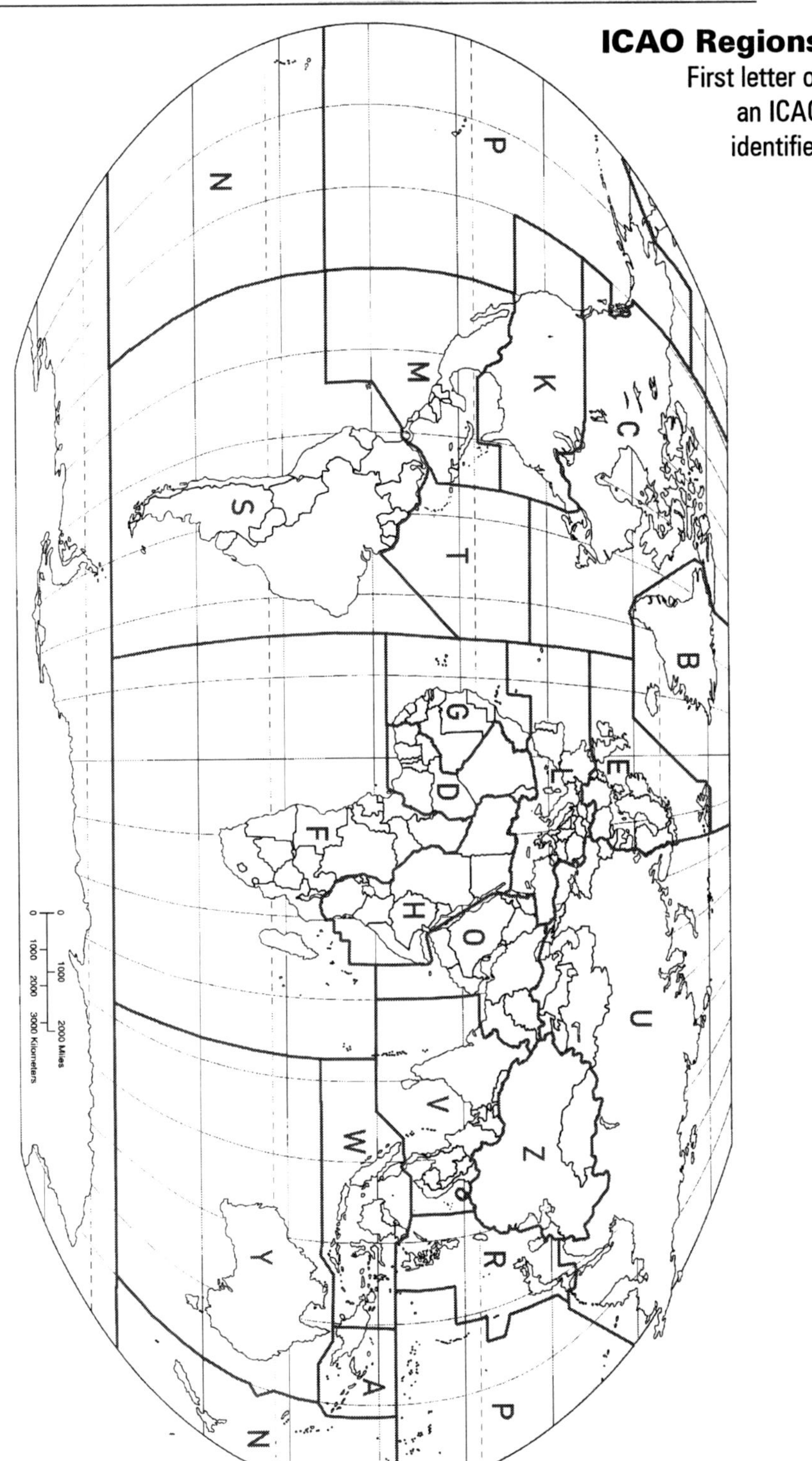

Reference

SECTION G

Administrative

G1. ADMINISTRATIVE

NWS Forecast Products

1. Routine forecasts

■ **Zone Forecast Product** (ZFP). Provides a public forecast for an area with sufficient climatological and meteorological homogeneity (typically two or three counties). It is issued at least four times per day (0430-0630L, 1030-1130L, 1530-1730L, and 2100-220L). (WSOM C-11)

■ **Local Forecast Product** (LFP). Provides a public forecast for a single town or city. It is issued at least four times per day (0430-0630L, 1030-1130L, 1530-1730L, and 2100-220L). (WSOM C-11)

■ **State Forecast Product** (SFP). Provides a general public forecast of hydrometeorological conditions over the forecast area for the next 5 days. It is issued twice daily, during the flexible window period allowed for early morning and early evening zone forecast releases. (WSOM C-10)

2. Hazard forecast categories

■ **Watch**. A watch is issued when conditions are favorable for a hazard but the risk is uncertain. This is used for the second and occasionally the third forecast period. (WSOM C-44)

■ **Warning**. Used when a non-precipitation hazard is occurring, very likely, or imminent, and is likely to pose a threat to life or property (above the criteria for an advisory). Exact criteria are listed below in each category. Generally this is issued only for the first period of a forecast but can be extended into the second period if confidence is high that the event will continue. (WSOM C-44)

■ **Advisory**. Used when a non-precipitation hazard is occurring, very likely, or imminent, and poses only a significant inconvenience (not meeting the criteria for a warning). Exact criteria are listed below in each category. Generally this is issued only for the first period of a forecast but can be extended into the second period if confidence is high that the event will continue. (WSOM C-44)

■ **Outlook**. Used when there is a good chance of a non-precipitation hazard occurring beyond the third or fourth period of the forecast. (WSOM C-44)

3. Hazard forecast types

■ **High wind** (NPW). These are issued when high winds are expected and a severe local storm, hurricane, or winter storm product is not appropriate. An **advisory** is issued for *sustained winds exceeding 25 mph*. A **warning** is issued for *sustained wind speeds of 40 mph or greater lasting for 1 hour or more*; or for *winds of 58 mph or more for any duration.* These criteria may be adjusted in local standard operating procedures. (WSOM C-44)

■ **Excessive heat** (NPW). An **advisory** is issued when two consecutive days are expected with a daytime *heat index reaching 105°F or more* with nighttime *minimum temperatures not falling below 80°F*. A **warning** (NPN) is issued in an extreme condition. (WSOM C-44)

■ **Fog** (NPW). Dense fog with visibility down to 1/4 mile is expected. (WSOM C-44)

■ **Winter storm** (WSW). Whether an advisory or warning is issued follows the subjective definitions provided above. (WSOM C-42)

■ **Heavy snow** (WSW). An **advisory** is generally issued when 1-3 inches of snow are expected, where a **warning** is issued for 4 inches. (WSOM C-42)

■ **Blizzard** (WSW). The following conditions are expected to prevail for 3 hours or more: sustained wind or frequent gusts to 35 mph or more, and considerable falling and/or blowing snow, frequently reducing visibility to less than 1/4 mile. (WSOM C-42)

■ **Freezing rain / freezing drizzle** (WSW). A hazard that occurs when rain or drizzle freezes on surfaces. This may produce risks by pulling down trees and utility lines. An **advisory** is generally issued when precipitation is light and the ice does not form on all exposed surfaces; a **warning** is indicated when the precipitation is expected to be heavier. (WSOM C-42)

■ **Ice storm** (WSW). Damaging accumulations of ice (typically 0.25 inch or more) are expected during freezing rain situations. Loss of power and communications throughout the forecast area are expected. (WSOM C-42)

■ **Sleet** (WSW). Ice pellets composed of frozen raindrops. (WSOM C-42)

■ **Freeze** (NPW). A condition when the surface air temperature is expected to be 32°F or lower over a widespread area for a climatologically significant period of time, and frost is not expected (due to wind or other factors). A **watch** or **advisory** cannot be issued; only a **warning** may be issued. (WSOM C-42)

■ **Frost** (NPW). A condition where ice crystals form on the ground during the growing season. It typically occurs when the surface air temperature is 32°F or lower and significant radiational cooling occurs. A **watch** is not issued. An **advisory** or **warning** may be issued depending on local policy. (WSOM C-42)

■ **Wind chill** (NPW). Generally the threshold for potentially dangerous wind chill is -20°F. However guidelines are specific to each forecast office. (WSOM C-42)

ADMINISTRATIVE

NWS Zones

NATIONAL WEATHER SERVICE

Alaska

Forecast Area Boundaries with forecast office identifier overprint
Public Zone Boundaries with three-digit zone number overprint
NWSI 10-507, Effective May 2005

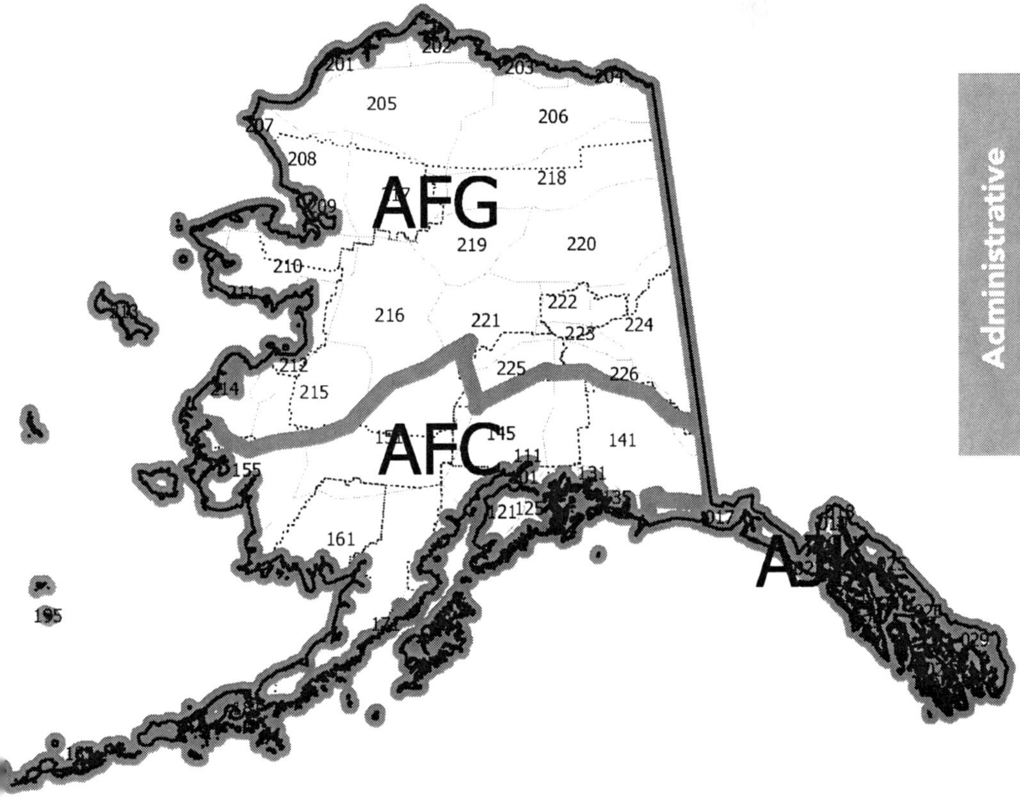

NATIONAL WEATHER SERVICE

Pacific Northwest

Forecast Area Boundaries with forecast office identifier overprint
Public Zone Boundaries with three-digit zone number overprint
NWSI 10-507, Effective May 2005

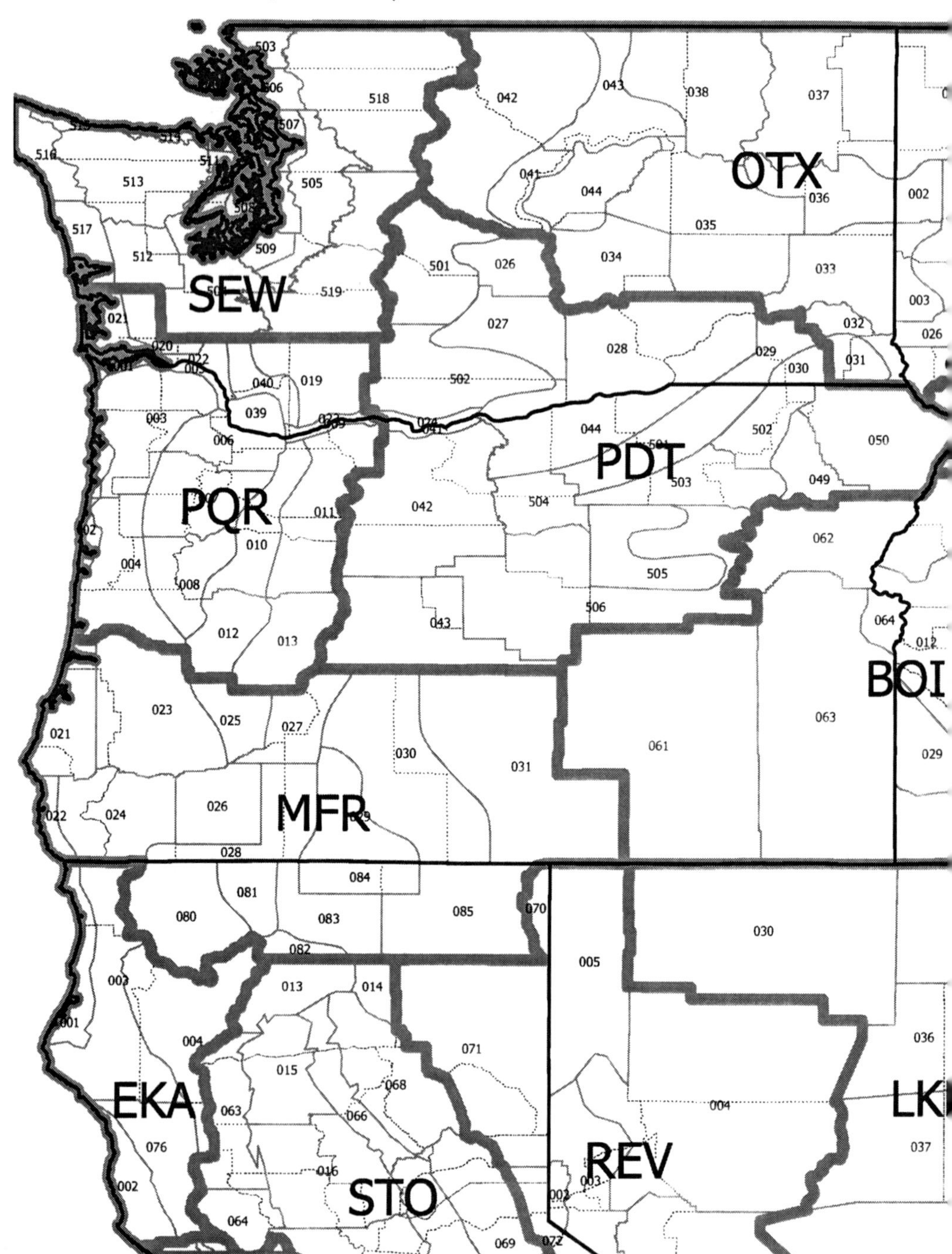

NATIONAL WEATHER SERVICE

Northern Rockies

Forecast Area Boundaries with forecast office identifier overprint
Public Zone Boundaries with three-digit zone number overprint
NWSI 10-507, Effective May 2005

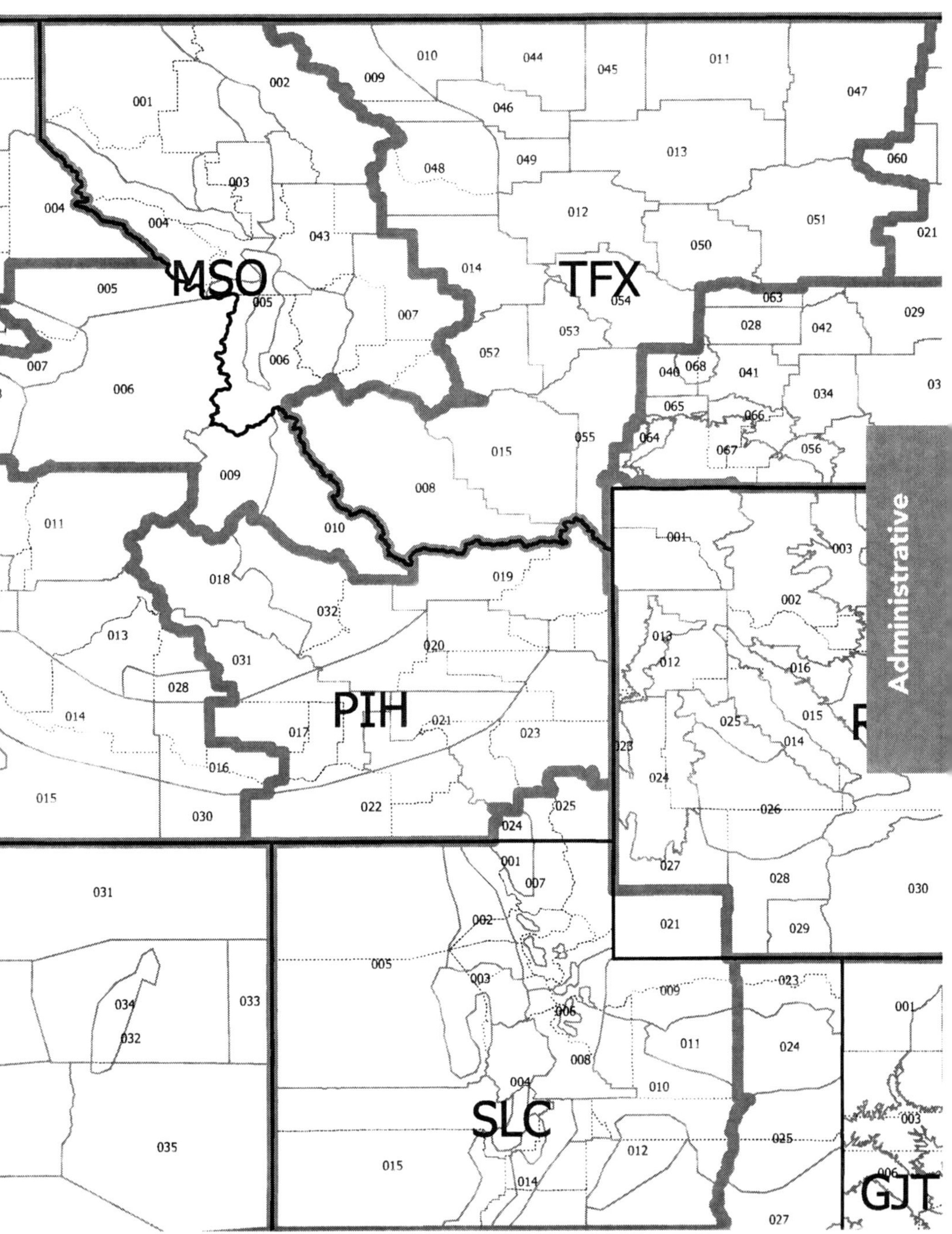

NATIONAL WEATHER SERVICE

Northern High Plains

Forecast Area Boundaries with forecast office identifier overprint
Public Zone Boundaries with three-digit zone number overprint
NWSI 10-507, Effective May 2005

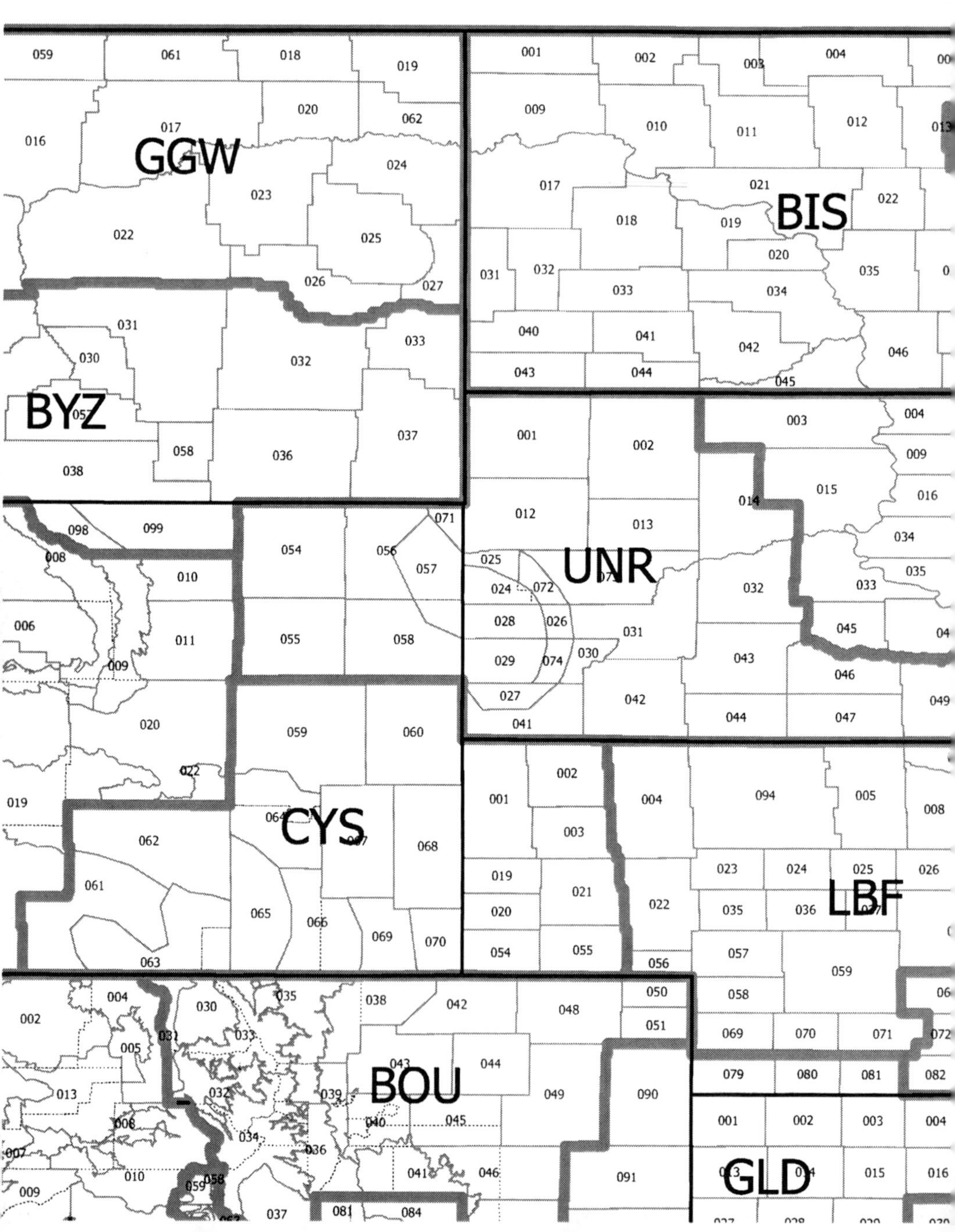

NATIONAL WEATHER SERVICE

Northern Low Plains

Forecast Area Boundaries with forecast office identifier overprint
Public Zone Boundaries with three-digit zone number overprint
NWSI 10-507, Effective May 2005

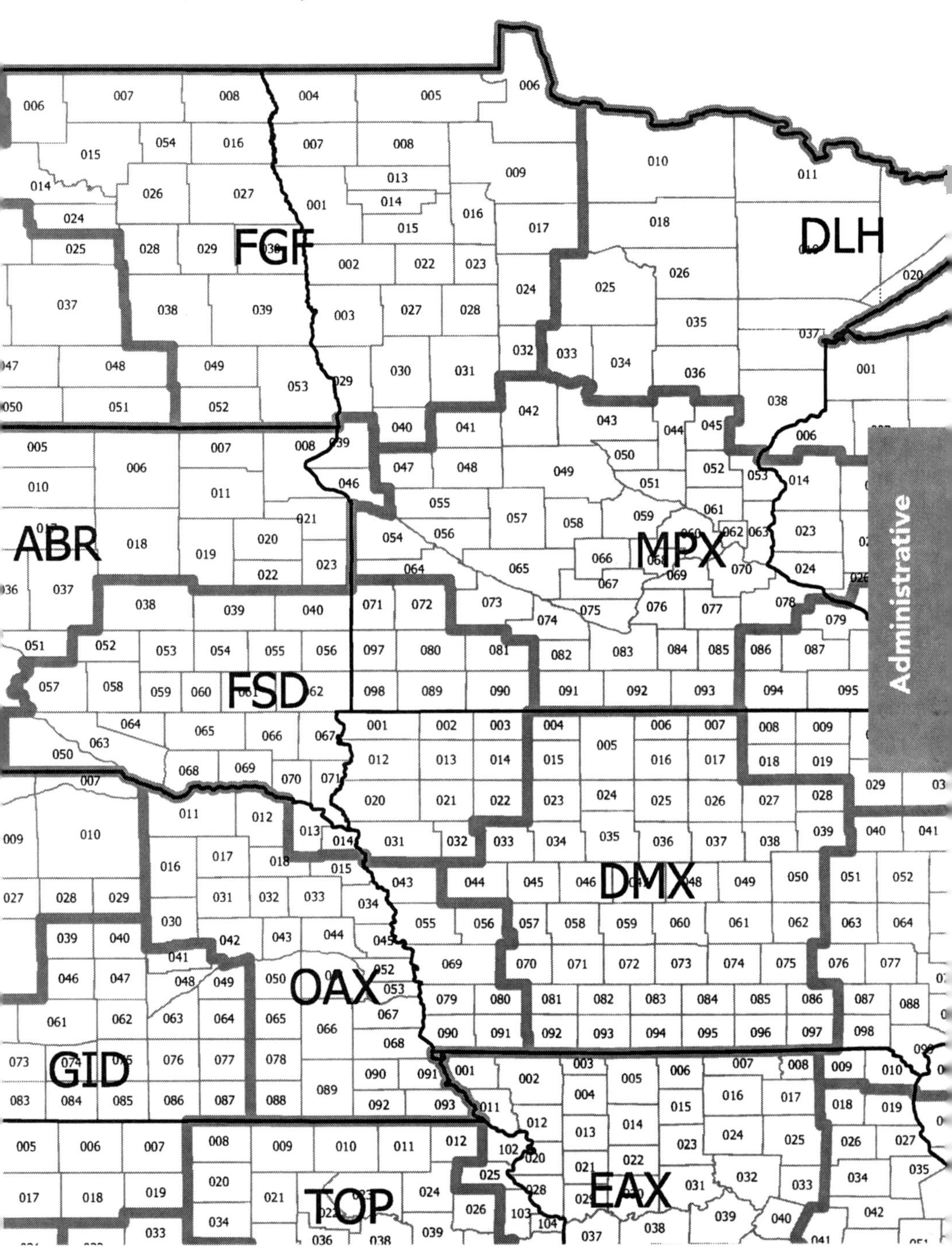

NATIONAL WEATHER SERVICE

Midwest

Forecast Area Boundaries with forecast office identifier overprint
Public Zone Boundaries with three-digit zone number overprint
NWSI 10-507, Effective May 2005

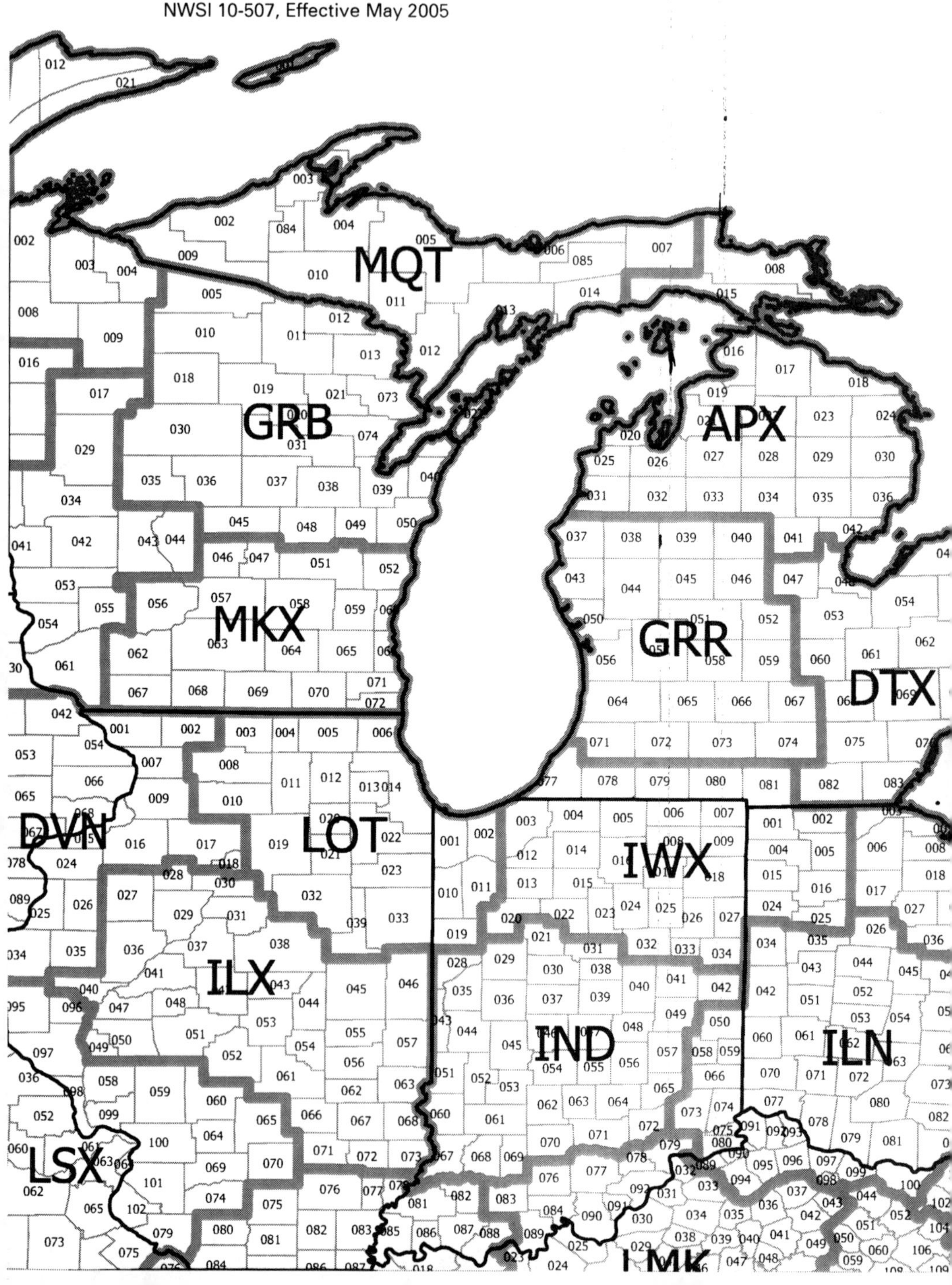

NATIONAL WEATHER SERVICE

Great Lakes

Forecast Area Boundaries with forecast office identifier overprint
Public Zone Boundaries with three-digit zone number overprint
NWSI 10-507, Effective May 2005

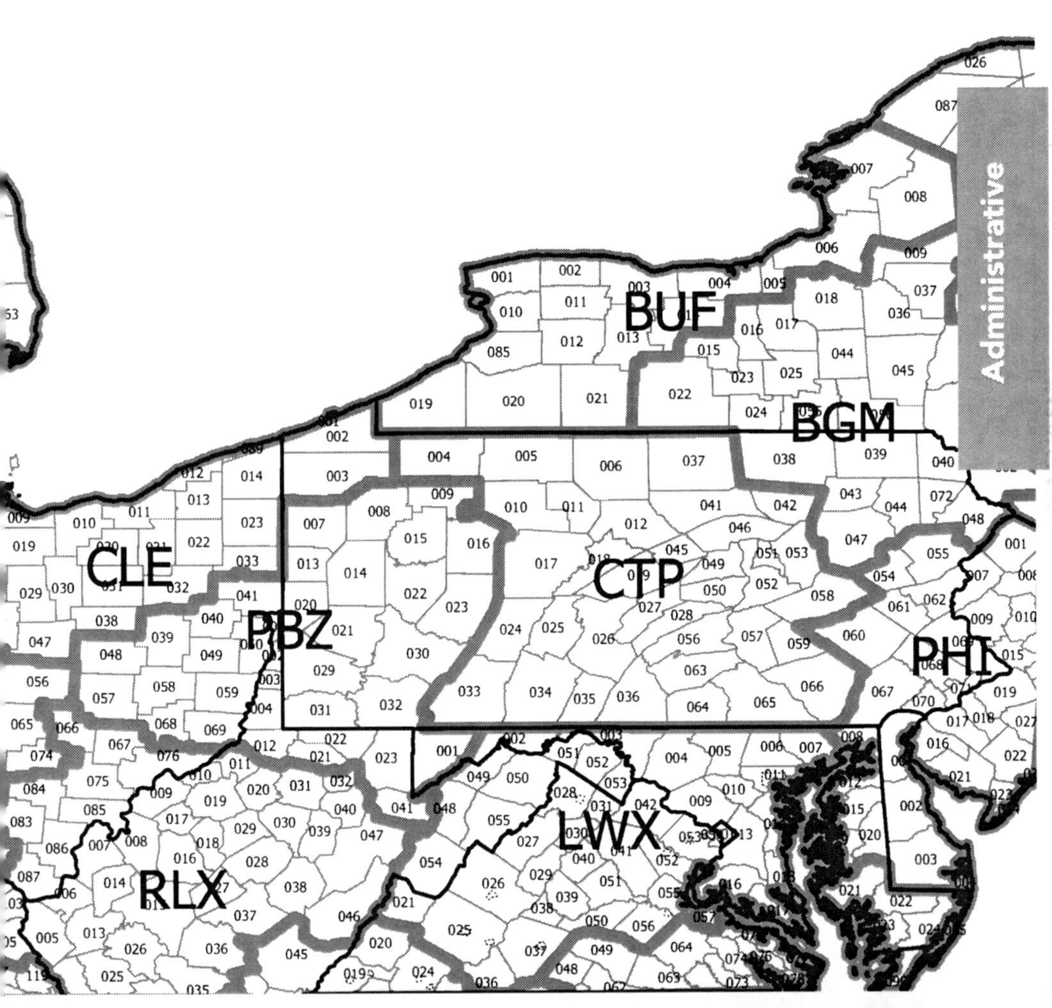

NATIONAL WEATHER SERVICE

New England

Forecast Area Boundaries with forecast office identifier overprint
Public Zone Boundaries with three-digit zone number overprint
NWSI 10-507, Effective May 2005

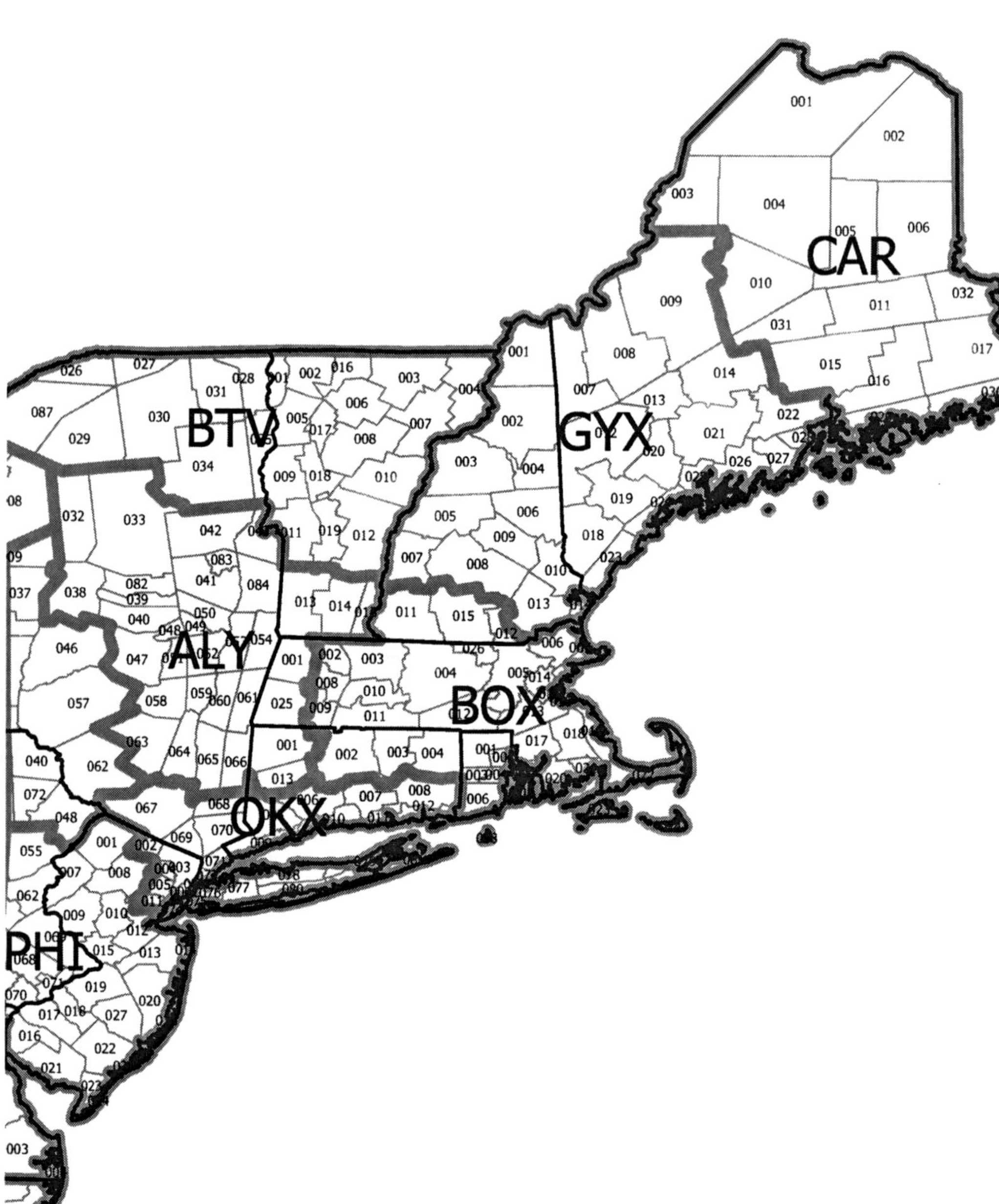

NATIONAL WEATHER SERVICE

Southern California

Forecast Area Boundaries with forecast office identifier overprint
Public Zone Boundaries with three-digit zone number overprint
NWSI 10-507, Effective May 2005

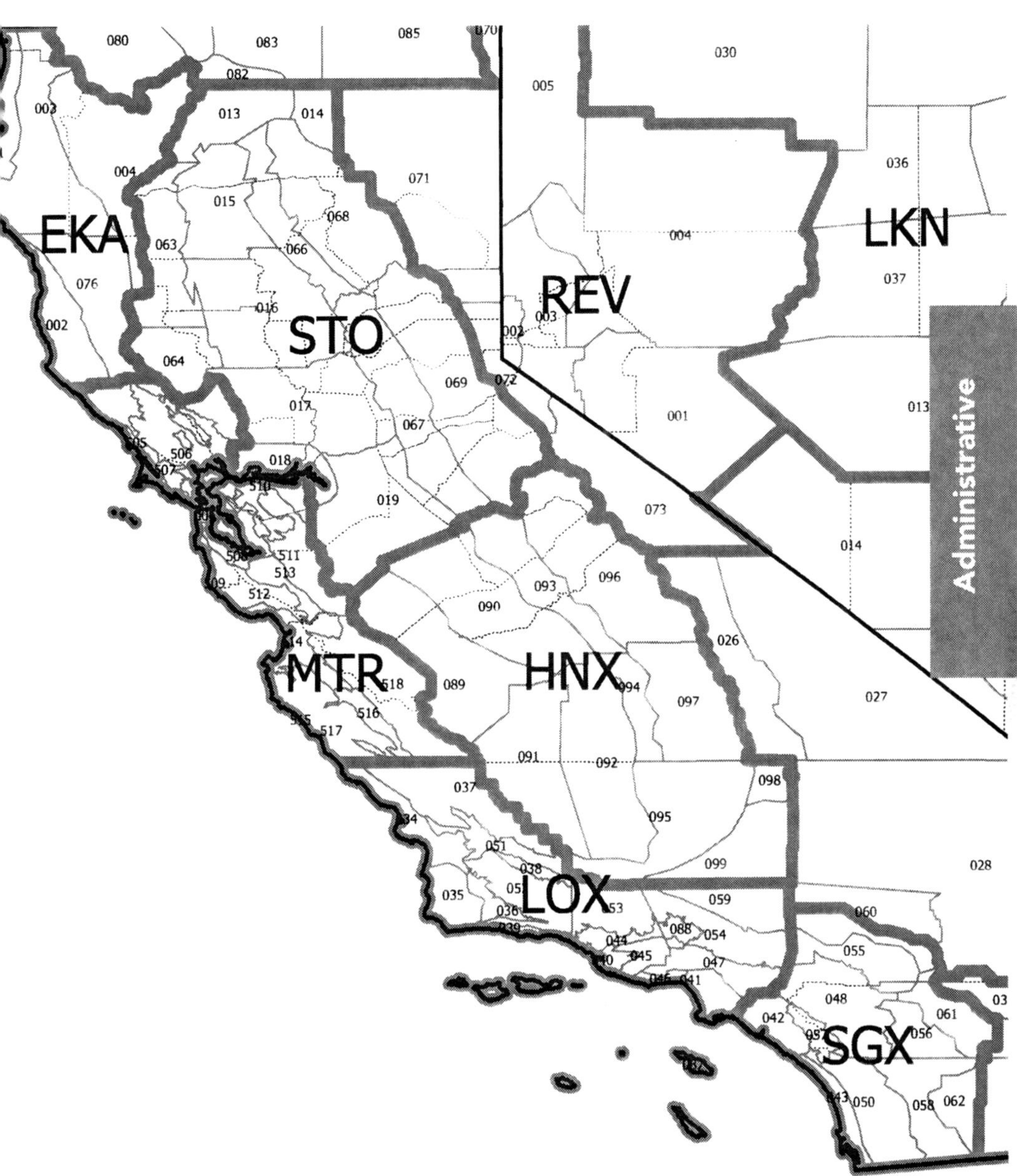

NATIONAL WEATHER SERVICE

Desert Southwest

Forecast Area Boundaries with forecast office identifier overprint
Public Zone Boundaries with three-digit zone number overprint
NWSI 10-507, Effective May 2005

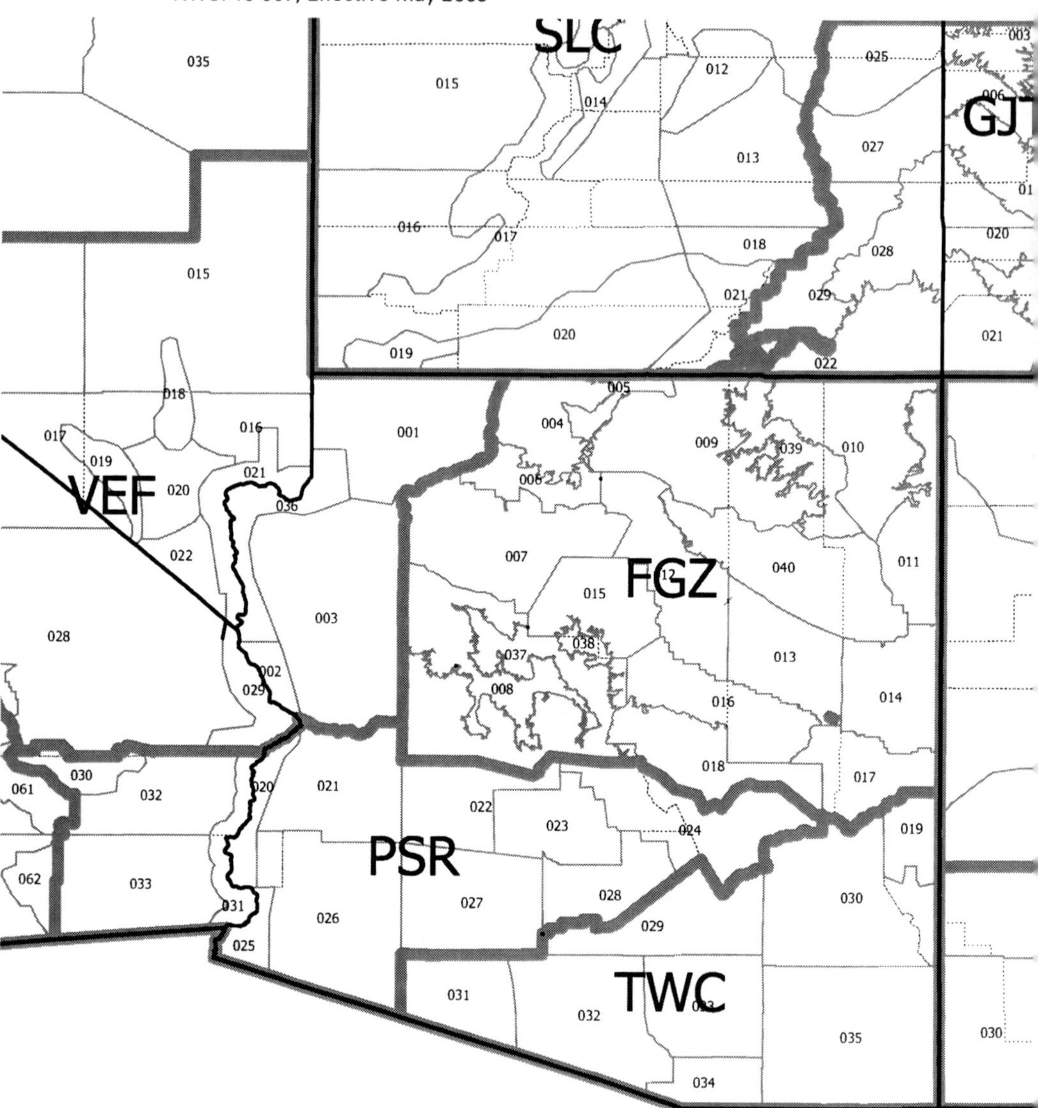

NATIONAL WEATHER SERVICE

Southern Rockies

Forecast Area Boundaries with forecast office identifier overprint
Public Zone Boundaries with three-digit zone number overprint
NWSI 10-507, Effective May 2005

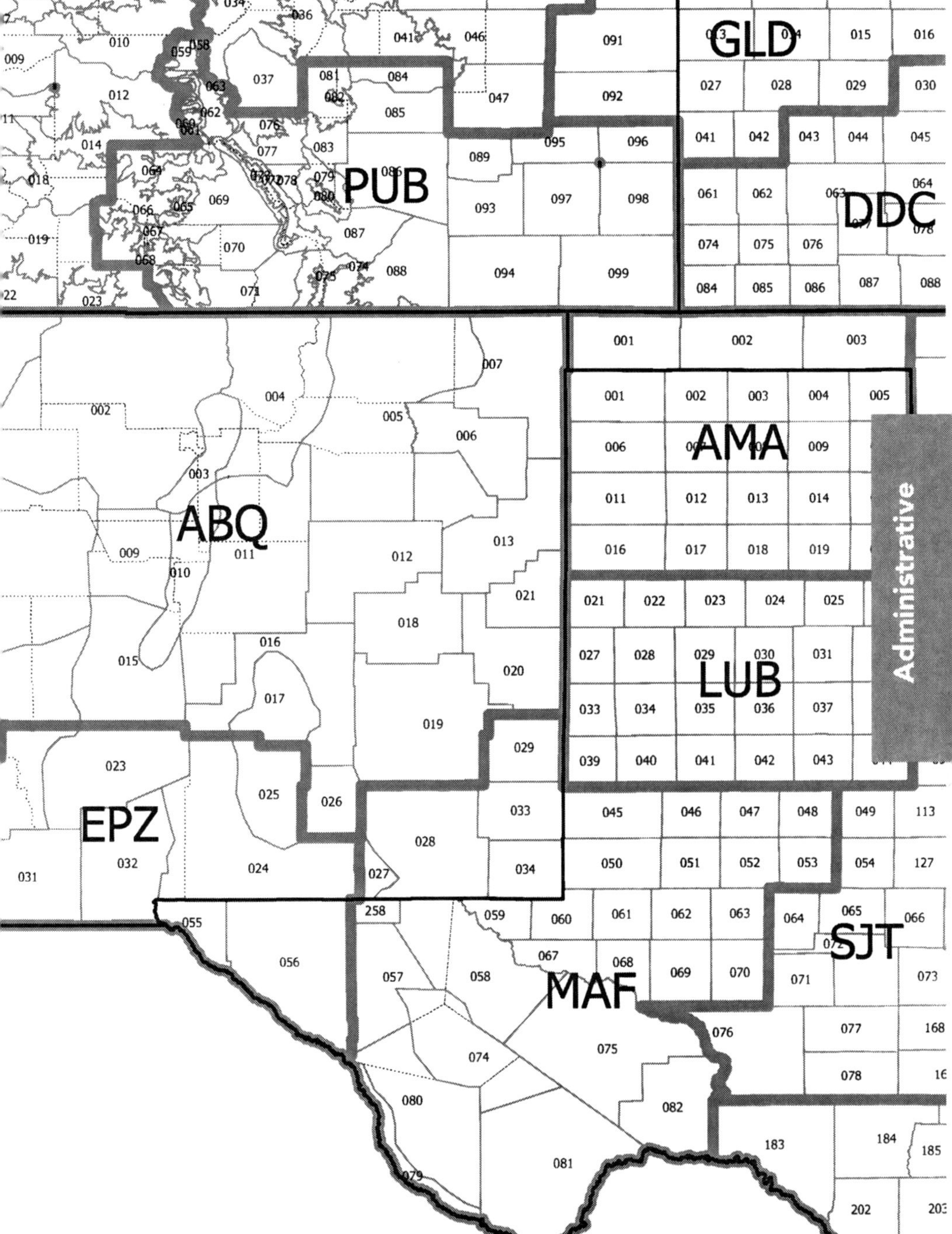

NATIONAL WEATHER SERVICE

Southern Plains

Forecast Area Boundaries with forecast office identifier overprint
Public Zone Boundaries with three-digit zone number overprint
NWSI 10-507, Effective May 2005

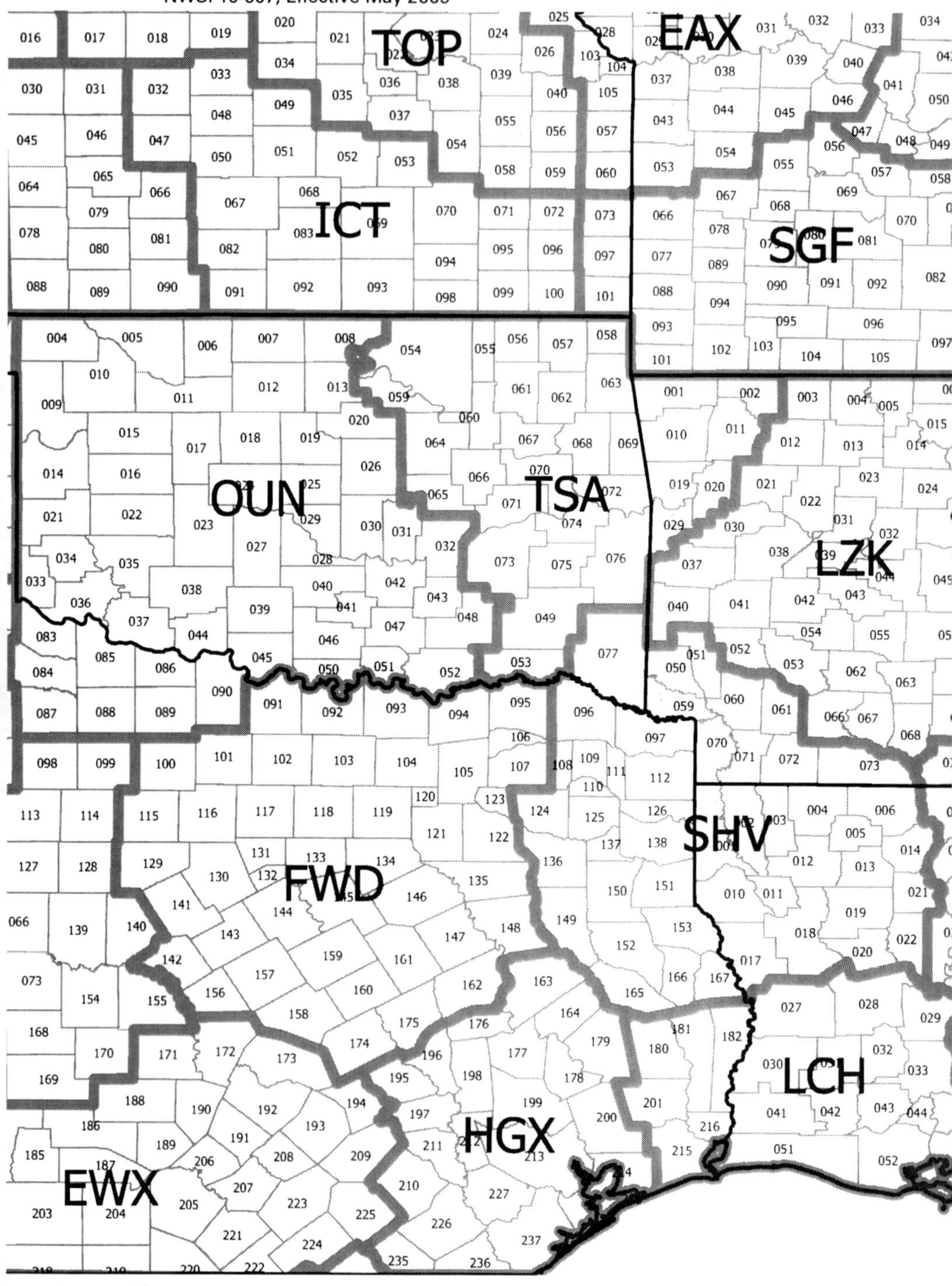

NATIONAL WEATHER SERVICE

Deep South

Forecast Area Boundaries with forecast office identifier overprint
Public Zone Boundaries with three-digit zone number overprint
NWSI 10-507, Effective May 2005

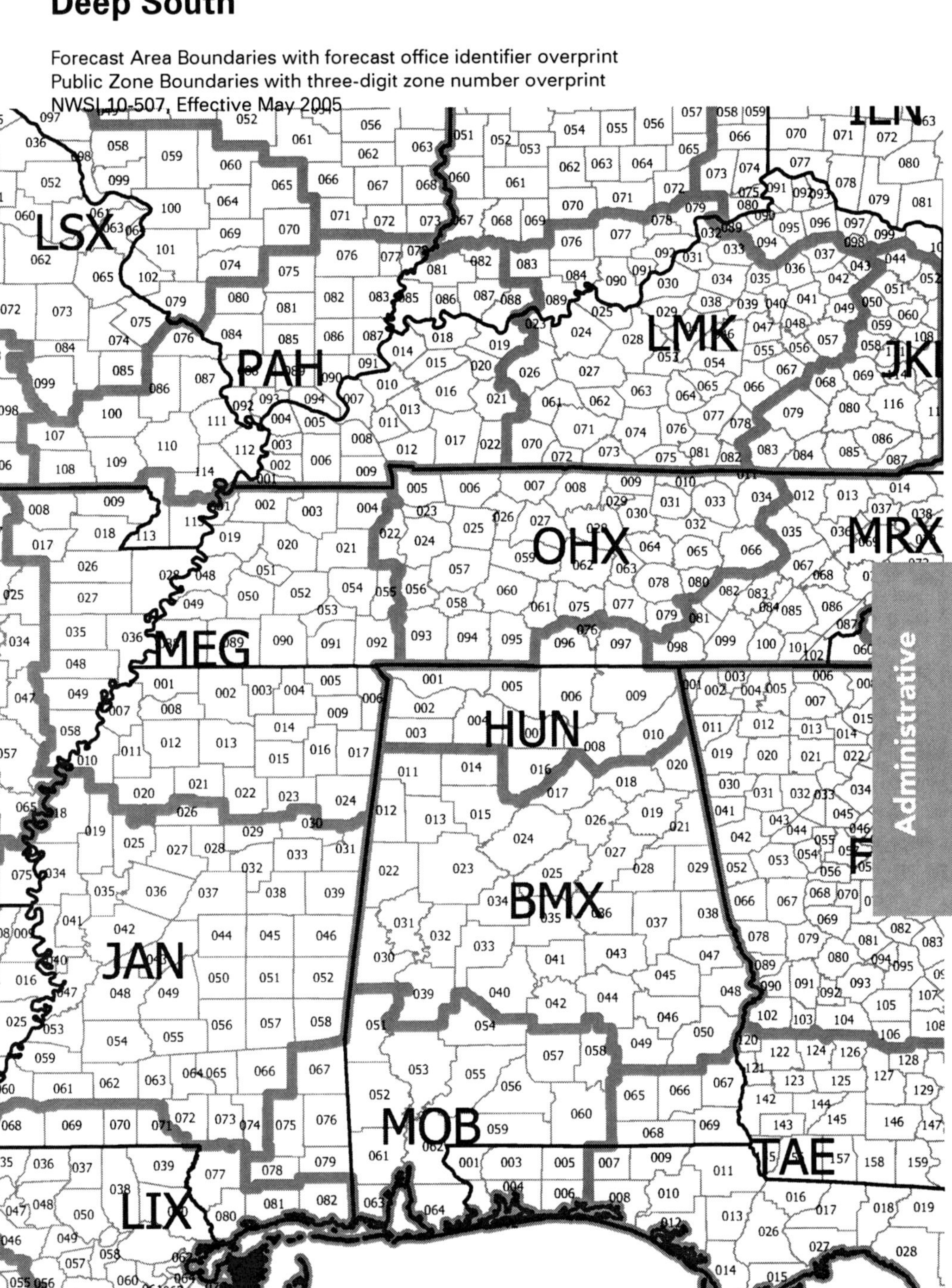

NATIONAL WEATHER SERVICE

Atlantic Seaboard

Forecast Area Boundaries with forecast office identifier overprint
Public Zone Boundaries with three-digit zone number overprint
NWSI 10-507, Effective May 2005

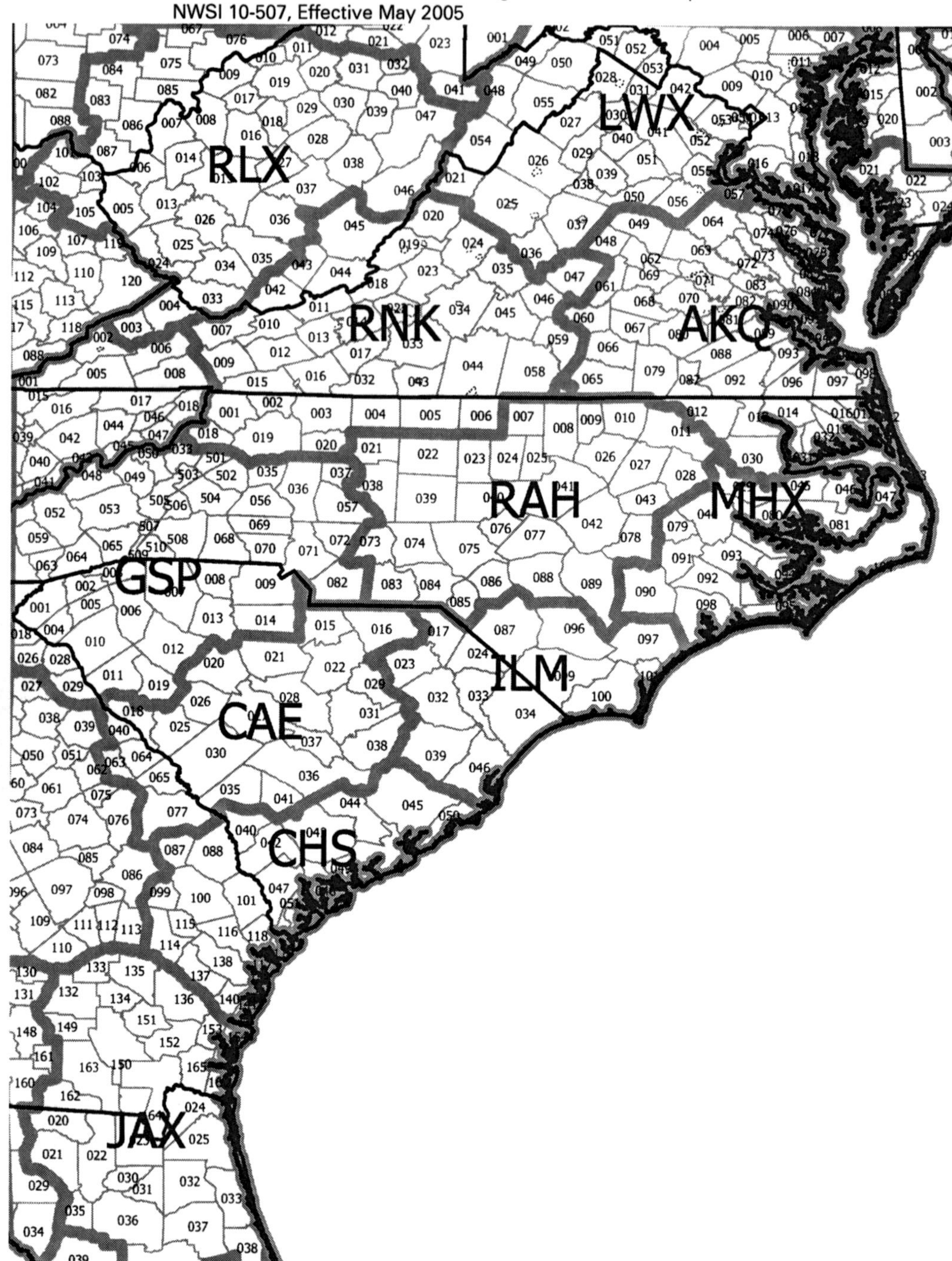

NATIONAL WEATHER SERVICE

Florida

Forecast Area Boundaries with forecast office identifier overprint
Public Zone Boundaries with three-digit zone number overprint
NWSI 10-507, Effective May 2005

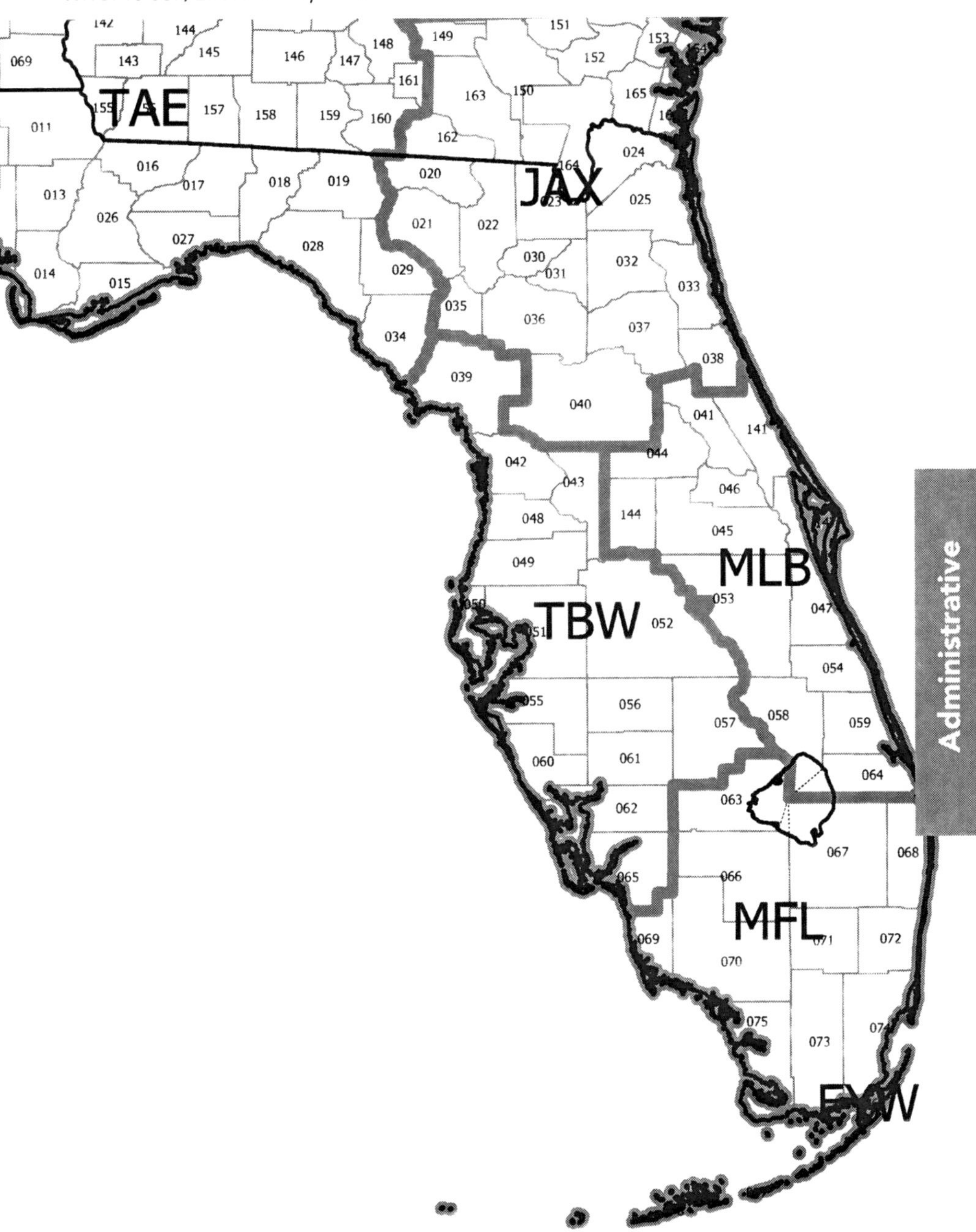

NATIONAL WEATHER SERVICE

South Texas

Forecast Area Boundaries with forecast office identifier overprint
Public Zone Boundaries with three-digit zone number overprint
NWSI 10-507, Effective May 2005

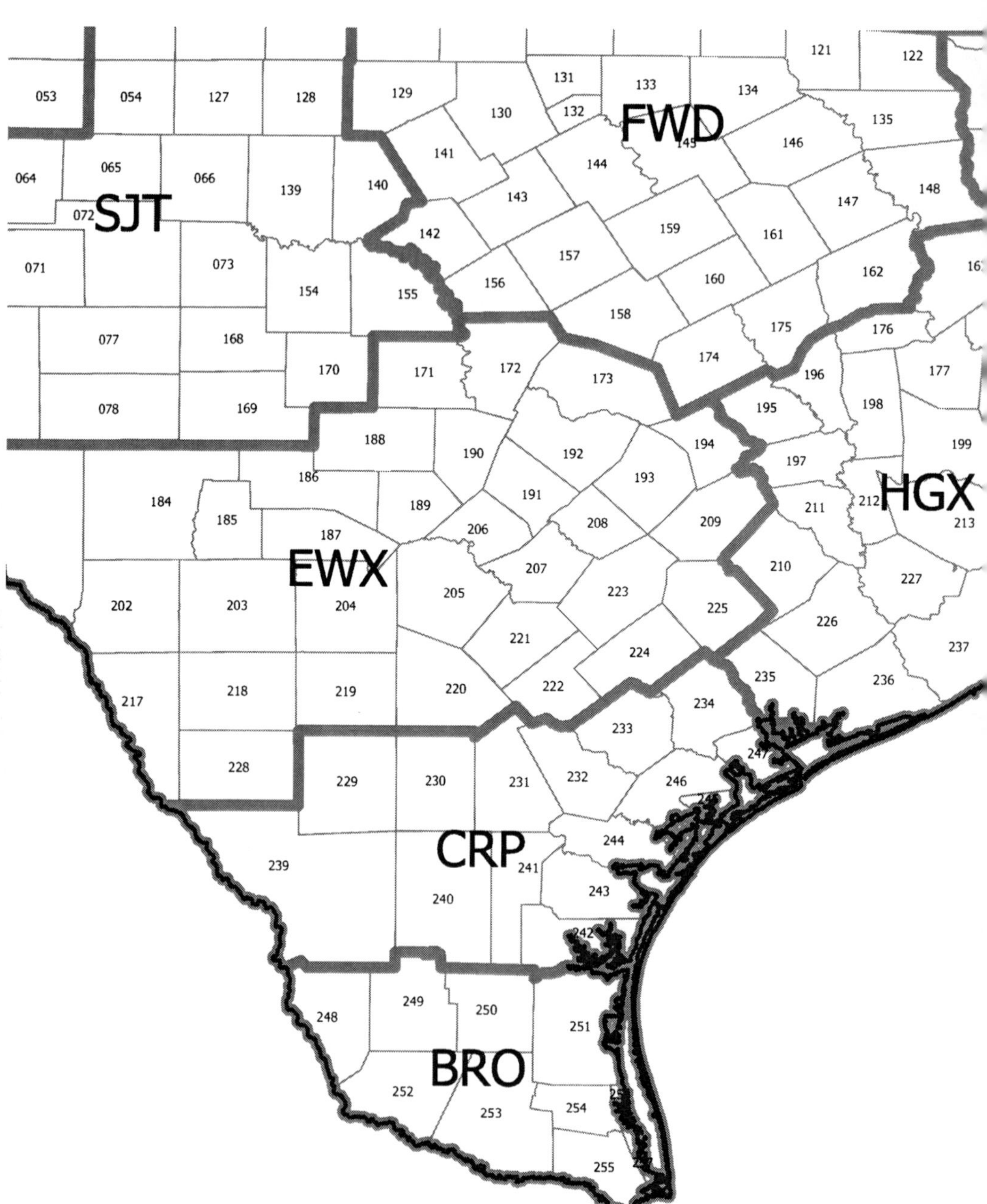

NATIONAL WEATHER SERVICE

Hawaii

Forecast Area Boundaries with forecast office identifier overprint
Public Zone Boundaries with three-digit zone number overprint
NWSI 10-507, Effective May 2005

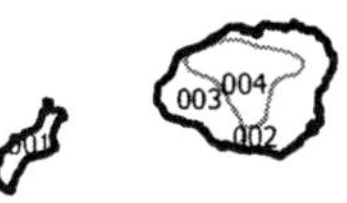

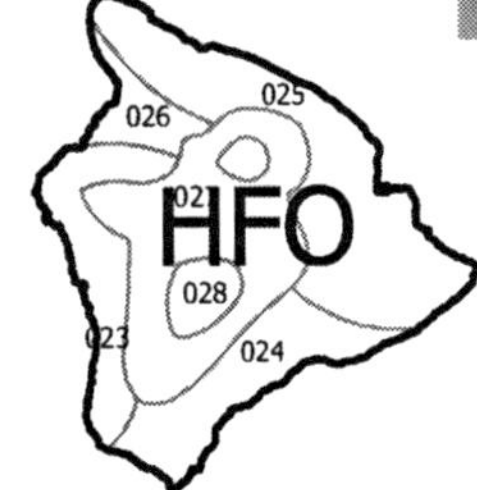

NATIONAL WEATHER SERVICE

Puerto Rico

Forecast Area Boundaries with forecast office identifier overprint
Public Zone Boundaries with three-digit zone number overprint
NWSI 10-507, Effective May 2005

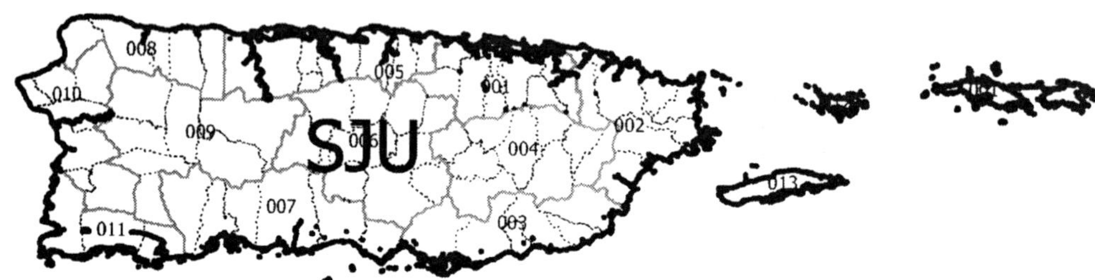

SECTION H

Glossary

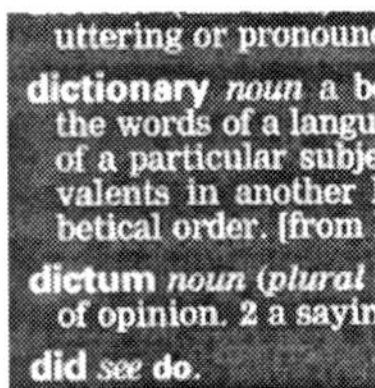

Glossary

REFERENCES

Contractions. Publication 7340.1Y. Federal Aviation Administration, 2005.

Forecasters Handbook #1: Centrally Produced Guidance and Analysis Products, National Weather Service, 1993.

Transport Canada Aeronautical Information Manual. Transport Canada, 2005.

A

A 1 altimeter (preceding four digits). 2 Absolute temperature. 3 Alaskan Standard Time. 4 arctic air mass

AO1 automated observation without precipitation discriminator

AO2 automated observation with precipitation discriminator

AAWF auxiliary aviation weather facility

ablation Depletion of snow and ice by melting and evaporation.

ABV above.

Ac Altocumulus (q.v.)

AC 1 Convective Outlook bulletin; from the Family of Services data stream header. 2 Altocumulus.

ACARS Aircraft Communications and Reporting System. A block of data transmitted by aircraft that often contains weather data.

ACC Altocumulus castellanus. [ICAO]

ACCAS Altocumulus castellanus. Altocumulus which forms in a convectively unstable layer.

accessory cloud A cloud which is dependent on a larger cloud system for development.

acre-foot The amount of water required to cover one acre to one foot of depth. This equals 326,851 gallons or 43,560 cubic feet.

ACRS across.

ACSL Altocumulus standing lenticular (q.v.)

ACT active.

ACYC Anticyclonic.

adiabatic The change in temperature without a transfer of heat. It may be caused by compression or expansion.

ADRNDCK Adirondack Mountains.

ADV 1 Advise. 2 Advection.

ADVCTN Advection.

advection Horizontal movement of air that causes changes in the physical properties of air at a specific location.

advection fog Fog that forms as warmer, moist air moves over a cold surface. The air is forced to condense as it loses heat to conduction.

advisory In the United States, a weather bulletin which is less serious than a warning.

ADVY Advisory.

AFC Area Forecast Center.

AFD Area Forecast Discussion (q.v.)

AFDK After dark.

AFGWC Air Force Global Weather Central, which became obsolete on 15 October 1997. It is now known as AFWA.

AFOS Automation of Field Operations and Services (discontinued). The backbone computer system of NWS offices; developed in 1976 and fielded in 1979; retired between 1996 and 1999.

AFT after.

AFL above freezing layer.

AFWA Air Force Weather Agency (q.v.)

AGL above ground level.

AHD ahead.

Air Force Weather Agency (AFWA) Weather component of the U.S. Air Force. Activated 15 October 1997, it is a combination of Air Weather Service Headquarters and the Air Force Global Weather Center.

air mass A body of air which contains relatively uniform properties of temperature and moisture.

AIRMS air mass.

AJ Arctic jet (q.v.)

albedo The portion of incoming radiation which is reflected by a surface.

ALF aloft.

ALGHNY Allegheny.

algorithm A computer program designed to solve a specific problem. Often used in WSR-88D radars.

aliasing A process in which a radar return has a frequency too high to be analyzed within the given sampling interval but at a frequency less than the Nyquist interval.

ALQDS All quadrants (all directions).

ALSEC All sectors.

ALSTG Altimeter setting (q.v.)

ALTA Alberta.

altimeter setting The pressure at which an altimeter must be set so that it reads the correct elevation.

altocumulus Mid-level clouds composed primarily of water or supercooled water. The base is traditionally at a height between 6,500 and 23,000 ft AGL (26,000 ft in the tropics and 13,000 ft at the poles).

Altocumulus Standing Lenticular (ACSL) Clouds formed at the tips of vertical waves in the wake of mountain ranges.

altostratus A bluish veil or layer of clouds having a fibrous appearance. The outline of the sun may show dimly as if through frosted glass. The base is traditionally at a height between 6,500 and 23,000 ft AGL (26,000 ft in the tropics and 13,000 ft at the poles).

ALUTN Aleutians.

ALWF Actual wind factor.

AM Ante meridian (a.m.)

AMD Amendment.

AMPLTD Amplitude.

AMS 1 American Meteorological Society. 2 Air mass.

anafront Also called *active front.* A cold front in which there is a tendency for air to ascend the frontal surface. Generally associated with lift and weather behind the front. (cf. *katafront*)

anemometer A device that measures wind speed.

ANL Analysis

ANLYS Analysis

anomalous propagation Unexpected radio wave propagation that occurs due to non-standard atmospheric conditions. Usually refers to ducting of the beam to the ground, returning ground clutter.

anticyclone An area of high pressure, around which the wind blows clockwise (counterclockwise in the Southern Hemisphere).

anvil The spreading top of a cumulonimbus cloud.

AOA At or above

AOB At or below

AP Anomalous propagation (q.v.)

APCH approach.

APCHG approaching.

APLCN Appalachian Mountains; Appalachians.

arctic air Air which has its roots over the snow-covered region of northern Canada, the polar basin, and northern Siberia.

arctic jet Baroclinic jet which develops in association with the polar vortex.

Area Forecast Discussion (AFD) A discussion of the meteorological thinking used in the creation of a zone forecast. (NWS)

ARINC Aeronautical Radio, Incorporated. Company based in Annapolis, MD responsible for ACARS (q.v.)

ARTCC Air Route Traffic Control Center

AS Altostratus (q.v.)

ASCII American Standard Code for Information Interchange

ASL above mean sea level.

ASOS Automated Surface Observing System. The network in place across the United States which have provided automated meteorological reports since 1992.

AST 1 Alaskan Standard Time. 2 Atlantic Standard Time.

ATLC Atlantic.

ATTM At this time

AURBO Aurora borealis.

AUTO Automated.

AVA Anticyclonic vorticity advection

AVN 1. The NCEP Aviation model, also known as the global spectral model, comprising the United States' primary global weather model; replaced by the GFS model (q.v.) 2. Aviation.

AWC Aviation Weather Center.

AWIPS Advanced Weather Interactive Processing System.

AWOS Automated Weather Observing System.

AWP Aviation Weather Processors.

B

B 1 Time of beginning of precipitation (suffixed with two digits for minute value). 2 Bering Standard Time.

back door cold front A cold front moving south or southwestward along and near the Appalachians.

backing Referring to a change in wind direction that is counterclockwise, with respect to either height or time. Contrast with *veering*.

backscatter Power that returns to the radar dish after striking a target.

BACLIN Baroclinic prognosis.

baroclinic zone An area in which a horizontal temperature gradient exists. Rapid weather changes may occur in such zones.

barotropic More properly referred to as *equivalent barotropic*, this term refers to a weather system which has weak or insignificant temperature contrasts

barotropic zone An area in which a significant horizontal temperature gradient does not exist. Rapid weather changes are not as likely as in a baroclinic zone.

Base Reflectivity (BR) A simple reflectivity product as obtained from any elevation of a radar scan (not necessarily the lowest one).

base velocity A simple velocity product as obtained from any elevation of a radar scan (not necessarily the lowest one).

BATROP Barotropic prognosis.

BC 1 Patches. 2 British Columbia.

BCFG patchy fog.

BCH beach.

BCKG backing.

BCOMG becoming.

BDA bermuda.

BDRY boundary ("BNDRY" is preferred).

beam width In radar meteorology, the width within which the power density is at least half that of the axis of the beam (i.e. within 3 dB)

BECMG becoming.

BFDK before dark.

BFR before.

BGN begin.

BGNG beginning.

BHND behind.

BINOVC Breaks in overcast.

BKN Broken cloud layer.

BL 1 Blowing. 2 Between layers.

BLDU Blowing dust.

BLDUP Buildup.

blizzard A winter storm which produces, for at least 3 hours, both winds gusting to 35 mph and falling/drifting/blowing snow reducing visibility to less than 1/4 mile.

BLKHLS Black Hills.

BLKT Blanket.

BLO 1 Below. 2 Below clouds.

block A long wave pattern, usually revealed on 200/250/300 mb charts, in which the long waves are neither progressing nor retrogressing. Often refers to the responsible feature, such as an omega or rex block (q.v.)

BLSA blowing sand.

BLSN blowing snow.

BLW below.

BLZD blizzard.

BNDRY boundary.

boundary layer (BL, PBL) The layer in contact with the ground in which friction is significant. This is usually the lowest few thousand feet of the atmosphere but may vary greatly with weather pattern, season, and insolation.

BOVC Base of overcast.

broken Partial coverage of the sky by a layer of more than half (5 to 7 oktas). (cf. *clear*, *few*, *scattered*, and *overcast*)

BR 1 base reflectivity (q.v.). 2 mist. [METAR]

BREF base reflectivity.

BRF 1 brief. 2 base reflectivity.

BRK Break(s).

BRKHIC Breaks in higher overcast.

BRKSHR Berkshire.

BRM Barometer.

BRN Bulk Richardson Number (q.v.)

BTL Between layers.

BTN between.

BTWN Between.

BUFR Binary Universal Format for Data Representation.

BULD Build.

Bulk Richardson Number (BRN) The ratio of CAPE to vertical wind shear. It has been found that values of less than 45 support supercellular structures, while greater than 45 favors multicells. However it is not as good of a predictor as its component terms are.

BWER Bounded Weak Echo Region.

BY Blowing spray.

C

C 1 Continental air mass (usually not capitalized). 2 Central Standard Time.

CA Clear air above.

CAA Cold air advection (q.v.)

CAN Canada.

cap A layer of warm air aloft that acts as an inversion and suppresses convective development. It may be measured by the Convective Inhibition Index, or CINH (q.v.)

CAPE Convective Available Potential Energy (q.v.)

CARIB Caribbean.

CASCDS Cascade Mountains.

CAT clear air turbulence.

CAVOK Ceiling and visibility okay.

CAVU Clear or scattered clouds and visibility unrestricted.

CAWS Common aviation weather subsystem.

CB, Cb Cumulonimbus (q.v.)

CBMAM Cumulonimbus mammatus.

CC, Cc Cirrocumulus (q.v.)

CCL Convective Condensation Level (q.v.)

CCLKWS Counter-clockwise.

CCSL Cirrocumulus standing lenticular.

CDFNT Cold front.

CDT Central Daylight Time.

CFP Cold front passage.

CHARC Characteristic.

CHC Chance.

chemtrail Contrail (q.v.) from which conspiracy theorists claim poisons are dispersed.

CHSPK Chesapeake.

CI, Ci Cirrus (q.v.)

CIG ceiling.

CIN Convective inhibition

CISK Convective instability of the second kind.

cirrocumulus (Cc) A layer of high, fibrous clouds with convective cells. The cloud is made up entirely of ice crystals. Its bases are traditionally as low as 16,000 ft (20,000 ft in the tropics; 10,000 ft in polar regions).

cirrostratus (Cs) A thin layer of high, fibrous clouds without detail and often appearing as a sheet covering the sky. It is composed entirely of ice crystals. Its bases are traditionally as low as 16,000 ft (20,000 ft in the tropics; 10,000 ft in polar regions).

cirrus (Ci) A layer of high, fibrous clouds composed entirely of ice crystals. Its bases are traditionally as low as 16,000 ft (20,000 ft in the tropics; 10,000 ft in polar regions).

CLD cloud.

clear Complete absence of cloud. (cf. *few, scattered, broken*, and *overcast*)

cloud height The height of a cloud's base, usually rounded to the nearest hundred feet (thousand feet above 10,000 ft).

CLR clear.

CLRG clearing.

CLRS clear and smooth.

CNCL cancel.

CNDN canadian.

CNTR center.

CNTRD centered.

CNVTV convective.

cold front The leading edge of an air mass that is replacing a warmer air mass.

Composite Reflectivity (CR) A WSR-88D radar product that displays the maximum reflectivity observed in a grid box at a given location.

CONDS conditions.

CONFDC Confidence.

confluence A pattern in which wind flows inward into a common axis. It is not the same as convergence. (cf. *difluence, convergence, divergence*)

CONT Continuous; continuously; continuity.

CONTDVD Continental Divide.

contrail Trail of condensation left by a high-flying jet in very cold, humid air.

CONTUS continuous.

CONUS conterminous United States.

convection The transport of heat and moisture by the vertical movement of air in an unstable atmosphere. This may cause cumuliform clouds and thunderstorms.

Convective Available Potential Energy (CAPE) the vertically integrated buoyancy of a rising air parcel. Measured in j/kg.

Convective Condensation Level (CCL) The height at which a parcel of air, if heated sufficiently from below, will rise adiabatically until saturation begins.

Convective Inhibition (CIN) A measure of negative buoyancy that prevents a rising parcel from reaching its Level of Free Convection, or LFC (q.v.). It is measured in j/kg.

convective temperature The theoretical surface temperature for a given atmospheric profile that must be reached to start the formation of convective clouds.

convergence A wind pattern in which more air is entering than leaving, either through speed convergence or confluence. (cf. *divergence, diffluence, confluence)*

CONUS Continental United States

Coordinated Universal Time (UTC) See Universal Coordinated Time.

COOSAC Committee on Operations, Standards, and Conventions

COR Correction.

Coriolis effect The effect caused by the Earth's rotation which deflects parcels to the right (left in the Southern Hemisphere).

COTR Contract Office Technical Representative

COTRAILS contrails.

couplet Adjacent maxima of radial velocities of opposite signs.

CPC Climate Prediction Center

CR Composite Reflectivity (q.v.)

cross section A diagram of the atmosphere in which horizontal distance is expressed on the X-axis and height on the Y-axis.

Cross Totals index (CT) an expression of instability, equalling Td_{850}-T_{500}. Values of greater than 18-30 are considered significant.

CS, Cs cirrostratus (q.v.)

CSI conditional symmetric instability

CST 1 coast. 2 Central Standard Time.

CT Cross Totals index (q.v.)

CTGY category.

CU, Cu cumulus (q.v.)

cumulonimbus (Cb) a large, cauliflower-shaped cloud whose upper portions are usually fibrous. Often associated with precipitation and thunder.

cumulus (Cu) low, heaplike clouds that are associated with convective weather. The three "categories" of cumulus are typically fair-weather, moderate, and towering. Further cumulus development will evolve into cumulonimbus.

CVA cyclonic vorticity advection.

cyclogenesis The intensification of a low-pressure system.

cyclone An area of low pressure with a closed circulation. The wind flow rotates counterclockwise (clockwise in the Southern Hemisphere).

D

DABRK daybreak.

dBZ Decibels of reflectivity factor.

DCAVU with remainder of report missing.

decoupling The intensification of the contrast between the boundary layer and the free atmosphere, which strengthens winds above and weakens winds below. Tends to occur at night.

DECRG decreasing.

DEG degree.

DELMARVA Delaware-Maryland-Virginia

derecho A widespread and fast-moving convective windstorm.

DFUS diffuse.

DIF 1 difference. 2 diffluence.

difluence Alternate spelling of *diffluence* (q.v.)

diffluence A pattern in which wind flows outward from a common axis. It is not the same as divergence. (cf. *confluence, convergence, divergence*)

DIST distant; distance.

diurnal 1 occurring on a daily basis. 2 occurring during the day only. (cf. *nocturnal)*

divergence a wind pattern in which more air is leaving than entering, either through speed divergence or diffluence. (cf. *convergence, diffluence, confluence)*

DKTS Dakotas.

DMSH diminish.

DMSHG diminishing.

DNS dense.

DNSLP downslope.

DNSTRM downstream.

DOCBLOCK computer program documentation block

DP 1 deep. 2 dewpoint.

DPNG deepening.

DPTH depth.

DR low drifting [METAR].

DRFT drift.

DRSA drifting sand [METAR].

DRSN drifting snow [METAR].

dryline a boundary which separates dry, warm continental air from moist, warm oceanic air. It is most common in the Great Plains, the Sahel, India/Bangladesh, Australia, and China.

DRZL drizzle.

DS dust storm. [METAR]

DSIPT dissipate.

DTLN International Date Line.

DTRT deteriorate.

DU widespread dust. [METAR]

DURC during climb. [PIREP]

DURD during descent. [METAR]

DURG during.

DVLPG developing.

DVV downward vertical velocity.

DWNDFTS downdrafts.

DWPT dewpoint.

DZ drizzle. [METAR]

dynamics A term that generally refers to forces produced by air out of geostrophic balance which in turn produces vertical motion.

E

E **1** east. **2** end of precipitation (followed by two-digit time vale in minutes). **3** Eastern Standard Time. **4** Equatorial air mass.

easterly wave A disturbance embedded in the trade winds that moves east to west.

EBDIC Extended Binary-Coded Decimal Interchange Code

ECMWF European Centers for Medium Range Weather Forecasting

EDT Eastern Daylight Time.

EHI Energy Helicity Index (q.v.)

EL Equilibrium Level (q.v.)

ELNGT elongate.

ELSW elsewhere.

ELY easterly.

EMBD embed.

EMBDD embedded.

EMC Environmental Modelling Center (q.v.)

EMSU Environmental Meteorological Support Unit.

ENDG ending.

Energy Helicity Index (EHI) An index that is a product of shear and instability, and is defined as CAPE x SRH / 160,000, where CAPE is j/kg and SRH is in m^2/s^2.

ENERN east-northeastern.

ENEWD east-northeastward.

ENTR **1** entire. **2** enter.

Environmental Modelling Center (EMC) A center of NCEP that is focused on improving numerical modelling technologies.

Equilibrium Level (EL) The height, sometimes within the

Glossary

stratosphere, at which a rising parcel's temperature becomes equal to that of the environment. Upward momentum is sharply lost beyond this point.

EOF 1 Expected Operations Forecast. 2 End Of File marker.

ESERN East-southeastern.

ESEWD East-southeastward.

EST Eastern Standard Time.

ETA Eta model

European Model The ECMWF global spectral model.

Exit region The region downstream from a jet max. The poleward side typically is associated with divergence aloft and upward motion; the equatorward side with convergence aloft and downward motion.

EXT External.

EXTTRAP Extrapolate.

EXTRM Extreme.

F

FA Area forecast.

FAA Federal Aviation Administration.

FAH Fahrenheit.

Family of Services (FOS) The public connection to National Weather Service data which was established in 1983.

FC Funnel cloud. [METAR]

+FC Tornado or waterspout. [METAR]

FCST forecast.

FD winds and temperatures aloft forecast

FEW partial coverage by cloud material of a quarter or less (1 to 2 oktas). (cf. *clear, scattered, broken, overcast)*

FFG Flash flood guidance

FG Fog. [METAR]

FIBI filed but impractical to transmit.

FILG filling.

FINO weather report will not be filed for transmission.

FL 1 Flight Level (followed by value in hundreds of feet MSL). 2 flash advisory.

FLDST flood stage.

FLG falling.

FLRY flurry.

FLWD followed.

FLWG following.

FLWIS flood warning issued.

FNL final production run for a given cycle

FNMOC Fleet Numerical Oceanography Center

FNOC Fleet Numerical Oceanographic Center (obsolete; replaced by FNMOC)

FNT front.

FNTGNS frontogenesis.

FNTLYS frontolysis.

FORNN forenoon.

FOS Family Of Services (q.v.)

FQT frequent.

FR from.

FRMG forming.

FROIN frost on indicator.

FROPA frontal passage.
FROSFC frontal surface.
FRQ frequent.
FRST frost.
FRWF forecast wind factor.
FRZ freeze.
FRZLVL freezing level.
FRZN frozen.
FT 1 feet; foot. 2 terminal forecast (obsolete; now TAF).
FTP File Transfer Protocol
FTS Federal Telecommunications System.
FU smoke. [METAR]
FULYR smoke layer aloft.
FUOCTY smoke layer over city.
FWC Fleet Weather Center.
FZ freezing.
FZDZ freezing drizzle. [METAR]
FZFG freezing fog. [METAR]
FZRA freezing rain. [METAR]

G

G gust; usually followed by speed of gust.
GBL Global production run for a given cycle
GDAS Global Data Assimilation System production run for a given cycle
GDM Graphic Display Model
geostrophic wind The imaginary wind that would result from a balance of both pressure gradient force and the Coriolis effect.
GES Guess
GFS Global Forecast System (q.v.)
GICG glaze icing.
GLFALSK Gulf of Alaska.
GLFCAL Gulf of California.
GLFMEX Gulf of Mexico.
GLFSTLAWR Gulf of St. Lawrence.
Global Forecast System (GFS) The most sophisticated global spectral model currently used by the United States. It incorporates both the AVN and MRF models, whose names have been "retired".
GMT Greenwich Mean Time
GND ground.
GNDFG ground fog.
GOES Geostationary Operational Environmental Satellite. The United States' network of geostationary weather satellites poised at the Equator above the Western Hemisphere continuously since 1974.
GR hail greater than 0.25 inch in diameter. [METAR]
GRAD gradient.
GRADS Grid Analysis and Display System.
GRBKS Grand Banks.
GRDL Gradual.
GRIB Gridded Binary data
GS Small hail or snow pellets 0.25 inch or less in diameter. [METAR]
GST gust.
GSTS gusts.

GSTY gusty.

GTS Global Telecommunications System.

GV ground visibility.

H

H 1 high pressure. 2 haze. [Airways code]

H2 200 mb level

H3 300 mb level

H5 500 mb level

H7 700 mb level

H8 850 mb level

HADS Hydrometeorological Automated Data System

HCVIS high clouds visible.

HDFRZ hard freeze.

HDSVLY Hudson Valley.

HGT height.

HI 1 high. 2 heat index.

HIC Hydrologist In Charge

HIEAT highest temperature equalled for all time.

HIEFM highest temperature equalled for the month.

HIESE highest temperature equalled so early.

HIESL highest temperature equalled so late.

HIFOR high level forecast.

HITMP highest temperature.

HIXAT highest temperature exceeded for all time.

HIXFM highest temperature exceeded for month.

HIXSE highest temperature exceeded so early.

HIXSL highest temperature exceeded so late.

HLSTO hailstone.

HLTP hilltop.

HLYR haze layer.

HMT Hydrometeorological Technician

hodograph A polar coordinate graph showing the wind profile of the atmosphere at a given point, with respect to ground-relative azimuth and speed.

HPC Hydrometeorological Prediction Center (q.v.)

HR hour(s).

HST Hawaii Standard Time.

HURCN hurricane.

HUREP hurricane report.

hurricane A warm-core tropical system that has sustained surface winds exceeding 63 kt.

HVY heavy.

HX high index.

Hydrometeorological Prediction Center (HPC) A center of NCEP which is responsible for centralized forecasting functions of the National Weather Service.

HZ haze. [METAR]

I

IC 1 icing. 2 ice crystals. [METAR]

ICAO International Civil Aviation Organization

ICG icing.

ICGIC icing in clouds.

ICGIP icing in precipitation.

IMC Instrument Meteorological Conditions.

IMDT immediate.

IMSL International Mathematical and Statistical Library

INCRG increasing.

INDEF indefinite.

INLD inland.

instability An atmospheric state in which warm air is able to continue rising and accelerating.

INSTBY instability.

INTMT intermittent.

INTR interior.

INTRMTRGN intermountain region.

INTS intense.

INTSFY intensify.

inversion An increase in temperature with height, comprising a stable layer in the atmosphere. Vertical motion through the inversion is suppressed.

INVOF in vicinity of.

INVRN inversion.

IOVC in overcast.

IR 1 infrared. 2 ice on runway.

isallobar a line of equal atmospheric pressure change.

isentrope a line of equal potential temperature.

isentropic lift lift produced by motion of air along surfaces of constant potential temperature which slope upward relative to the parcel's motion. This typically occurs when the parcel is traversing from warmer to colder air below.

isentropic subsidence Sinking motion produced by motion of air along surfaces of constant potential temperature which slope downward relative to the parcel's motion. This typically occurs when the parcel is traversing from colder to warmer air below.

ISLD isolated.

isobar a line of equal pressure.

isochrone a line of equal time.

ISOLD isolated.

ISPAN Information Stream Project for AWIPS/NOAAPORT.

J

J/KG joules per kilogram

jet max A region of maximum winds within a jet stream. Also *jet streak, speed max*.

jet streak A region of maximum winds within a jet stream. Also *jet max, speed max*.

JIF Job Implementation Form

JMA Japan Meteorological Agency

JSC Johnson Spaceflight Center

JSPRO Joint Systems Program Office for NEXRAD

JTSTR jet stream.

K

K cold air mass.

katafront also called *inactive front*; a cold front in which there is a tendency for air to descend the frontal surface. Generally associated with subsidence behind the front and weather ahead of the front. (cf. *anafront*)

KFRST killing frost.

K-Index (KI) A measure of the thunderstorm potential based on vertical temperature lapse rate, moisture content of the lower atmosphere, and the vertical extent of the moist layer. Equals T_{850}-T_{500}+Td_{850}-DD_{700} where DD equals dewpoint depression. Values above 20-35 are significant.

KI K-Index (q.v.)

knot A measure of velocity, nautical miles per hour, equal to 1.15 statute miles per hour.

KT knot (q.v.)

L

LABRDR Labrador.

lapse rate The change in temperature with height. Normally is 6.5 Celsius degrees per km.

LAWRS Limited Aviation Weather Reporting Station (usually a control tower)

LCL **1** local. **2** lifted condensation level.

LCN Loosely Coupled Network

LCTMP little change in temperature.

LDG landing.

LFT lift.

LI Lifted Index (q.v.)

LCL Lifted Condensation Level (q.v.)

LEWP Line echo wave pattern

LFC Level of free convection

LFQ Left-front quadrant of a jet streak. In the Northern Hemisphere this is usually associated with upward motion.

LGRNG long range.

LGT light.

LIFR low instrument flight rules.

LK lake.

LLJ low-level jet.

LLWS low level wind shear; generally used only in context of aviation operations.

LN line.

LO low pressure area.

LOEAT lowest temperature equalled for all time.

LOEFM lowest temperature equalled for month.

LOESE lowest temperature equalled so early.

LOESL lowest temperature equalled so late.

LOTMP lowest temperature.

LOXAT lowest temperature exceeded for all time.

LOXFM lowest temperature exceeded for the month.

LOXSE lowest temperature exceeded so early.

LOXSL lowest temperature exceeded so late.

Lifted Condensation Level (LCL) The height at which a parcel of air will become saturated if lifted adiabatically.

Lifted Index (LI) The temperature difference between a lifted parcel and that of its environment at 500 mb. This is a single-level expression of instability. It equals T_E-T_P where E is the environment and P is the parcel. Negative values are unstable, and below -5 are significant.

LLJ Low Level Jet (q.v.)

Low Level Jet (LLJ) An elongated area of strong winds, generally below 10,000 ft MSL, which may occur in advance of extratropical lows. It is significant in transporting heat and moisture poleward, reinforcing baroclinicity and destabilizing the atmosphere.

long wave A large-scale wave in the upper atmosphere, either a trough or a ridge. There are usually four or five long waves around a hemisphere.

LRQ Left-rear quadrant of a jet streak. In the Northern Hemisphere this is usually associated with downward motion.

LSR loose snow on runway.

LTG lightning.

LTGCA lightning cloud-to-air.

LTGCC lightning cloud-to-cloud.

LTGCCCG lightning cloud-to-cloud and cloud-to-ground.

LTGCG lightning cloud-to-ground.

LTGCW lightning cloud-to-water.

LTGIC lightning in cloud.

LTL little.

LTLCG little change.

LTNG lightning (usually LTG).

LVL level.

LWIS Limited Weather Information System.

LWR lower.

LX low index.

LYR layer (of cloud).

M

M 1 missing. 2 minus zero (value usually follows). 3 Maritime air mass (usually not capitalized). 4 Mountain Standard Time. 5 in runway visual reports, indicates value lower than lowest sensor value.

M2/S2 Meters squared per second squared

MA map analysis.

MAN Manitoba.

MAR Modernization and Associated Restructuring Program

MAX Maximum

Maximum Parcel Level (MPL) The highest attainable level a thunderstorm updraft can reach, where all further upward velocity of a parcel is lost. Factors in overshooting tops.

MB Millibars

MCC Mesoscale convective complex

MCD Mesoscale discussion

MCIDAS Man-Computer Interactive Data Access System

MCS Mesoscale Convective System

MDFYD modified.

MDR Manually Digitized Radar (now obsolete)

MDT 1 moderate. 2 Mountain Daylight Time.

MEGG merging.

mesocyclone A low pressure area which is the embodiment of a rotating thunderstorm; it usually measures 1 to 5 miles in diameter. It is a misnomer because it is not a mesoscale system.

mesolow A mesoscale low-pressure area. Not to be confused with mesocyclone.

mesohigh A mesoscale high-pressure area, sometimes associated with stagnating thunderstorm outflow air.

mesoscale referring to weather systems with scales of about 50 to 500 miles, or 1 to 24 hours.

METAR Meteorological Aviation Report.

MEX Mexico.

MHKVLY Mohawk Valley.

MI shallow. [METAR]

MIC Meteorologist In Charge

MID middle.

MIFG shallow fog. [METAR]

MIFN midnight.

MLCAPE Mean Layer CAPE. Calculated using a parcel that contains mean temperature and mixing ratio of a layer, typically 100 mb deep.

MLTLVL melting level.

MMG main meteorological office.

MNLD mainland.

MOA Memorandum of Agreement

MOD 1 moderate. 2 modify.

Model Output Statistics (MOS) A statistical forecasting model, usually calculated city-by-city.

MOGR moderate or greater.

monsoon A seasonal shift in wind direction.

MONTR monitor.

MOS Model Output Statistics (q.v.)

MOU Memorandum of Understanding

MOV move.

MOVG moving.

MPH statute miles per hour.

MPL Maximum Parcel Level (q.v.)

MPC Marine Prediction Center

MRF Medium Range Forecast model (obsolete; replaced by GFS)

MRGL marginal.

MRNG morning.

MRTM maritime.

MS 1 Mississippi. 2 minus.

MSL (above) Mean Sea Level

MSLP Mean Sea Level Pressure
MST Mountain Standard Time.
MSTLY mostly.
MSTR moisture.
MTN mountain.
MTS mountains.
MUCAPE Most Unstable CAPE. CAPE calculated from a parcel that provides the most unstable CAPE possible.
MVFR marginal visual flight rules.
MX mixed icing.
MXD mixed.
MXG mixing.

N

N north.
NASA National Aeronautics and Space Administration
National Centers for Environmental Prediction (NCEP) Was NMC (National Meteorological Center) from 1958-1995. An agency falling under NOAA which provides guidance and products to the National Weather Service. It is comprised of nine centers: Aviation Weather Center (AWC); Climate Prediction Center (CPC); Environmental Modelling Center (EMC); Hydrometeorological Prediction Center (HPC); NCEP Central Operations (NCO); Ocean Prediction Center (OPC); Space Environmental Center (SEC); Storm Prediction Center (SPC); and Tropical Prediction Center (TPC).
National Climatic Data Center (NCDC) The United States government agency responsible for archival of meteorological data.
National Oceanic and Atmospheric Administration (NOAA) The United States government agency falling under the Department of Commerce, which is responsible for all civilian programs engaged in work with the atmosphere, oceans, and lakes.
National Weather Service (NWS) A branch of NOAA responsible for all United States public forecasting.
NB New Brunswick.
NCCF NOAA Central Computer Facility
NCDC National Climatic Data Center (q.v.)
NCEP National Centers for Environmental Prediction (q.v.)
NCO NCEP Central Operations
NCWX no change in weather.
NE **1** northeast. **2** no echoes on radar.
negative tilt Description of an upper-level trough whose axis is tilted to the west with increasing latitude. It is often associated with strengthening dynamics.
NELY northeasterly.
NERN northeastern.

NESDIS National Environmental Satellite Data and Information Service

NEXRAD Next Generation Weather Radar (WSR-88D)

NEXUS Next Generation Upper-Air System

NEW ENG New England.

NFLD Newfoundland.

NGM Nested Grid Model

NGT night.

NHC National Hurricane Center

NIDS NEXRAD Information Dissemination Service

nimbostratus an amorphous cloud thick enough to completely obscure the sun, with its base almost indistinguishable and typically obscured by precipitation. Does not produce showers or thunder. Abbreviated Ns.

NLY northerly.

NM nautical mile (6076 ft).

NMBR number.

NMC National Meteorological Center (obsolete; now NCEP)

NMFS National Marine Fisheries Service

NMRS numerous.

NNERN north-northeastern.

NNEWD north-northeastward.

NNWRN north-northwestern.

NNWWD north-northwestward.

NO not available.

NOAA National Oceanic and Atmospheric Administration (q.v.)

NORIP no pilot balloon observation unless weather changes significantly.

NOS National Ocean Survey

NOSPL no special observations.

NOTAM Notice to Airmen

nowcast a forecast of about six hours or less. Also called a *short-term forecast.*

NPRS nonpersistent.

NR near.

NRLY nearly.

NS, Ns **1** nimbostratus (q.v.) **2** Nova Scotia.

NSCSWD no small craft or storm warnings displayed.

NSSFC National Severe Storms Forecast Center

NSW no significant weather.

NVA Negative vorticity advection. Advection of negative vorticity into a region.

NW northwest.

NWLY northwesterly.

NWRN northwestern.

NWS National Weather Service (q.v.)

NWSTG National Weather Service Telecommunications Gateway

OBS observation.

OBSC obscured; obscuration.

occlusion The convergence of three air masses, in which the least dense is displaced aloft and the remaining two are demarcated by an *occluded front*. Typically occurs when a cold front "catches up" to a warm front.

OCFNT occluded front.

OCLD occlude.

OCLN occlusion.

OCNL occasional.

OFOAR Office of Oceanic and Atmospheric Research

OFP occluded front passage.

OFSHR offshore.

OHP One-Hour Precipitation, as used in weather radar estimates.

OI Optimum Interpolation method

okta An eighth of sky cover.

omega block An upper-level pattern in which a high pressure (height) area intensifies to a very high amplitude, resembling the greek letter omega. It "locks in" the long wave pattern.

OMTNS over mountains.

ON Office Note.

ONSHR onshore.

ONT Ontario.

ORGPHC orographic.

OSV ocean station vessel.

OTAS on top and smooth.

OTLK outlook.

OTP on top.

OTW otherwise.

OTWZ otherwise.

outflow boundary The leading edge of outflow from a thunderstorm downdraft. It may persist hours or days after the dissipation of the storm.

OV position is over the following location. [PIREP]

OVC overcast.

overcast A cloud layer completely covering the sky (8 oktas of cover). (cf. *clear*, *few*, *scattered*, and *broken*)

overrunning An oversimplification of the process of *isentropic lift* (q.v.).

OVR over.

OVRNG overrunning.

P

P 1 polar air mass. 2 Pacific Standard Time. 3 In a runway visual report, visibility is higher than the highest sensor value.

P6SM Visibility greater than 6 statute miles.

PAC Pacific.

PBL 1 planetary boundary layer. 2 probable.

PCPN precipitation.

PD period.

PDMT predominant.

PDT Pacific Daylight Time.

PDW priority delayed weather.

PE 1 ice pellets. [METAR] 2 Primitive Equation model.

PEN peninsula

PFJ Polar front jet (q.v.)

PGTSND Puget Sound.

PIBAL pilot balloon.

PIREP Pilot Report

PISE no pilot balloon due to sea conditions.

PISO no pilot balloon due to snow.

PIWI no pilot balloon due to high wind.

PK WND peak wind.

PL ice pellets. [METAR]

PLW snow plow.

PNHDL Panhandle.

PO dust whirls; sand whirls. [METAR]

polar front jet (PFJ) The jet that is associated with the gradient between polar and tropical air masses. (cf. *arctic jet, subtropical jet, low-level jet*)

polar vortex A large cold-core low aloft that typically is found over northern Hudson Bay in North America during the winter months. Occluding polar front systems are usually absorbed into the polar vortex.

POP Probability of Precipitation (q.v.)

positive area The area formed on a sounding between an environmental temperature line and a warmer parcel temperature line. Its area is roughly proportional to CAPE.

positive-tilt Description of an upper-level trough whose axis is tilted to the east with increasing latitude. It is often associated with weakening dynamics.

potential temperature The temperature which a parcel would have if brought to a common level, by standard convention 1000 mb.

PPINA plan position indicator (radar scope) not available.

PPINE plan position indicator (radar scope) shows no echoes.

PPINO plan position indicator (radar scope) not operational.

PPIOK plan position indicator (radar scope) okay/resumed.

PPIOM plan position indicator (radar scope) offline for maintenance.

PQ Quebec.

PR partial.

PRBLTY probability.

PRECDD preceded.

PRECDS precedes.

PRES pressure.

PRESFR pressure falling rapidly.

PRESRR pressure rising rapidly.

pressure gradient The change in pressure over a given distance.

PRFG partial fog. [METAR]

PRJMP pressure jump.

PROB*nn* chance is *nn* percent. [TAF]

Probability of Precipitation (POP) A quantity that describes the likelihood of a measurable amount of precipitation at any given location in a forecast area. The NWS expressions are 20% for slight chance, 30-50% for a chance, and 60-70% for likely.

PROD Production (for operational jobs)

profiler Also *wind profiler*. A radio detection device designed to measure wind direction and speed vertically in the troposphere above a given point.

PROG prognosis; prognostic.

PRSNT present.

PRSTG persisting.

PS plus.

PSG passage.

PSN position.

PSR packed snow on runway.

PST Pacific Standard Time.

PTCHY patchy.

PTLY partly.

PVA Positive vorticity advection. Equal to CVA in the Northerm Hemisphere and AVA in the Southern Hemisphere (q.v.)

Q

Q vector A horizontal vector representing the rate of change of the horizontal potential temperature gradient. Convergence or divergence of the vectors are associated with forcing for vertical motion.

QC Quebec.

QG Quasi-geostrophic

QPF Quantitative Precipitation Forecast

QSTNRY quasistationary.

QUE Quebec.

R

R 1 runway. 2 rain. [Airways code]

RA rain. [METAR]

RABA no rawinsonde observation due to balloons not available

RABAL rawinsonde balloon wind data

RABAR rawinsonde balloon release.

RACO no rawinsonde observation due to communications outage.

RADAT 1 freezing level height in feet. 2 radiosonde observation data.

RADNO report missing account radio failure.

RAOB Radiosonde observation

radial velocity The component of motion along an axis extending from a radar unit. The NEXRAD base velocity product depicts radial velocity.

RAFI radiosonde report not filed.

RAFRZ radiosonde observation freezing levels.

RAFS Regional Analysis and Forecast System (NGM)

RAHE no rawinsonde observation due to no helium available.

RAICG radiosonde observation icing follows.

range folding A process by which a radar echo returns after another pulse has been transmitted, creating an echo that might be incorrectly distanced by the radar unit.

RAOB radiosonde observation.

RAPI radiosonde observation already sent in PIBAL collection.

RAREP radar report

RASN rain and snow. [METAR]

RAVU Radiosonde Analysis and Verification Unit.

RAWE no rawinsonde observation due to unfavorable weather.

RAWI no rawinsonde observation due to strong or gusty winds.

RAWIN upper wind observation by radio methods.

RCD radar cloud detection report.

RCDNA radar cloud detection report not available.

RCDNE radar cloud detection report shows no echoes.

RCDNO radar cloud detection not operational.

RCDOM radar cloud detection offline for maintenace.

RCKY Rocky Mountains.

RCM Radar Coded Message. An automated product of the WSR-88D unit which provides a summary of the echoes and signatures from a given radar.

RDG ridge.

RDWND radar dome wing.

RESTR restrict; restricted.

rex block A blocking pattern in the upper levels in which a closed high is located poleward of a closed low. The long-wave pattern tends to "lock up".

RFQ Right-front quadrant of a jet streak. In the Northern Hemisphere this is usually associated with downward motion.

RFRMG reforming.

RGD ragged.

RGL Regional Model

RGN region.

RH relative humidity.

RHINO radar range height indicator not operational.

ridge An elongated area of high pressure or heights.

RMK Remark.

RMNG remaining.

RNFL Rainfall.

ROBEPS Radar operating below equipment performance standards.

RPD rapid.

RRQ Right-rear quadrant of a jet streak. In the Northern Hemisphere this is usually associated with upward motion.

RS record special observation; a scheduled observation meeting special criteria (in airways observations used until 1996).

RSG rising.

RUC Rapid Update Cycle model

RUF rough.

RUNHIST Run History

RVR Runway Visual Range.

RWY runway.

S

S south.

SA 1 sand. [METAR] 2 surface airways observation (used until 1996).

SASK Sasketchewan.

SBCAPE Surface based CAPE; resulting from a parcel that is lifted from the surface with no other modifications.

SBSD subside.

SC, Sc Stratocumulus (q.v.)

scattered Partial coverage of a cloud layer, covering more than a quarter to half of the sky, of 2 to 4 oktas. (cf. *clear*, *few*, *broken*, and *overcast*)

SCSL stratocumulus standing lenticular.

SCT scattered (q.v.)

SCTR sector.

SD **1.** Radar Report (now obsolete) **2.** *Storm Data*, a publication of NCDC.

SDM Senior Duty Meteorologist.

SE southeast.

SEC Space Environment Center

SELS Severe Local Storms Unit (used until 1995; see Storm Prediction Center).

SELY southeasterly.

SERN southeastern.

SEV severe. [ICAO; U.S. uses "SVR"]

sferics lightning detection by radio methods.

SG snow grains. [METAR]

SGD solar geophysical data.

SH showers. [METAR]

SHFT shift.

SHGR hail showers. [TAF]

SHLW shallow.

Showalter Stability Index (SSI) The difference in temperature between the environment at 500 mb and a parcel lifted from 850 mb, expressed as T_{500}-T_{850}. A negative value corresponds to instability. Lifted Index and CAPE are usually preferred over the SSI.

SHRA rain showers. [TAF]

SHSN snow showers. [TAF]

SHWR shower.

SIERNEV Sierra Nevada Mountains.

SIGMET Significant Weather bulletin for pilots

SIR snow and ice on runway.

SKC sky clear.

SLD solid.

SLO slow.

SLP sea level pressure.

SLR slush on runway.

SLT sleet.

SLY southerly.

SM statute mile (5280 ft).

SMK smoke.

SML small.

SMTH smooth.

SN snow. [METAR]

SNBNK snowbank.

SNFLK snow flake.

SNOINCR snow depth increase during past hour; value follows.

SNRS sunrise.

SNST sunset.

SNW snow.

SNWFL snowfall.

SP **1** station pressure. **2** special observation. [Airways code]

SOO Science and Operations Officer

sounding A plot of temperature and dewpoint above a given station with respect to temperature (X-axis) and height (Y-axis). A thermodynamic diagram, usually the SKEW-T log P, is typically used.

SPC Storm Prediction Center (q.v.)

SPECI special observation. [METAR]

spectrum width The variance in velocity of scatterers within a given volume of air.

speed max A region of maximum winds within a jet stream. Also *jet max, jet streak*.

SPENES NESDIS satellite precipitation estimate

SPKL sprinkle.

SPLNS southern Plains.

SPRD spread.

SQ wind squall. [METAR]

SQAL squall.

SQLN squall line.

SRH Storm-relative helicity

SS sandstorm. [METAR]

SSERN south-southeastern.

SSEWD south-southeastward.

SSI Showalter Stability Index (q.v.)

SSWRN south-southwestern.

SSWWD south-southwestward.

ST, St Stratus (q.v.)

STAGN stagnating; stagnation.

STBL stable.

STFRA stratus fractus.

STFRM stratiform.

STG strong.

STGTN strengthen.

STJ Subtropical jet (q.v.)

STK Storage Technology

STM storm.

STNRY stationary.

Storm Prediction Center (SPC) branch of NCEP, located in Norman, Oklahoma, which is responsible for providing short-term forecast guidance for convective storms.

stratocumulus A relatively flat, low cloud with little vertical development. It has distinct globular masses or rolls.

stratus (St) A low, sheetlike cloud which may either occur alone, or with precipitation (in which case it is referred to as *scud, fractus,* or *stratus of bad weather*).

subsidence Sinking motion.

sub-synoptic Mesoscale.

subtropical jet (STJ) An upper-level jet stream that is usually found between 20 and 30 deg of latitude and is associated with thermal differences within the subtropical high. (cf. *arctic jet, polar front jet,* and *low-level jet*)

SVR severe.

SVRL several.

SW southwest.

SWLG swelling.

SWLY southwesterly.

SWRN southwestern.

SWODY1 Severe Weather Outlook - Day 1

SWODY2 Severe Weather Outlook - Day 2

SX stability index.

SXN section.

SYNOP synoptic.

synoptic-scale Spanning a distance scale of over 500 miles or a time scale of days.

SYNS synopsis.

SYS system.

T

T 1 trace. 2 temperature. 3 tropical.

TA air temperature. [PIREP]

TAF Terminal Aerodrome Forecast.

TB turbulence. [PIREP]

TCU towering cumulus.

TD Tropical Depression

teleconnection A strong statistical relationship between weather in different parts of the globe.

TEMPO temporary changes expected.

THD thunderhead.

theta-e Equivalent potential temperature

THK thick.

THKNG thickening.

THN thin.

THNC thence.

THNG thinning.

THRU through.

THRUT throughout.

THSD thousand.

TIL until.

TKOF takeoff.

TM time. [PIREP]

TOP cloud top.

Total Totals Index (TTI) A sum of the Cross Totals and Vertical Totals indices. It is equal to T_{850}-T_{500}+Td_{850}-T_{500}. A value of greater than 44-56 is considered significant.

TOVC top of overcast.

TP aircraft type. [PIREP]

TPG topping.

TPB Technical Procedures Bulletin

TPC Tropical Prediction Center (q.v.)

TRIB tributary.

TRML thermal.

TROF trough.

TROP 1 tropopause. 2 tropical.

Tropical Prediction Center (TPC) A branch of NCEP responsible for tropical weather forecasting, including hurricanes.

tropical storm A warm-core storm with a maximum sustained surface wind of 34-63 kt.

tropopause The point between the troposphere and stratosphere at which a positive tropospheric lapse rate becomes neutral or negative.

trough An elongated area of low pressure or heights.

TROWAL trough of warm air aloft.

TRPCD continental tropical air mass.

TRPCL tropical.

TRPLYR trapping layer.

TRRN terrain.

TS 1 thunderstorm. [METAR] 2 tropical storm.

TSGR thunderstorm with hail. [METAR]

TSGS thunderstorm with small hail. [METAR]

TSHWR thundershower.

TSPL thunderstorm with ice pellets.

TSQLS thundersqualls.

TSRA thunderstorm with rain. [METAR]

TSSA thunderstorm with sand. [METAR]

TSSN thunderstorm with snow. [METAR]

TSTM thunderstorm.

TTI Total Totals Index (q.v.)

TURB turbulent.

TURBC turbulence.

TURBT turbulent.

TWD toward.

TWRG towering.

typhoon A tropical storm of hurricane strength in the Western Pacific basin.

U

UA routine pilot report

UAG upper-air geophysics.

UCAR University Corporation for Atmospheric Research

UCL UNICOS Control Language (shell script)

UDDF updrafts and downdrafts.

UKMO United Kingdom Met Office

UKMET United Kingdom Met Office

ULJ Upper level jet

UNSBL unseasonable.

UNSTBL unstable.

UNSTDY unsteady.

UNSTL unsettle.

UP unknown precipitation. [METAR]

UPR upper.

UPS Uninterruptable Power Supply

UPSLP upslope.

UPSTRM upstream.

UTC Universal Coordinated Time

UUA urgent pilot report.

UVV upward vertical velocity.

UWNDS upper winds.

V

V variable; varies.

VA volcanic ash. [METAR]

VAD Velocity Azimuth Display. A plot of radial velocity (Y-axis) with respect to azimuth (X-axis) by a weather radar for a given level. It is used as a basis for construction of VWP diagrams (q.v.)

VAFTAD Volcanic Ash Forecast Transport and Dispersion.

VC **1** vicinity. **2** see AK-47.

VCFG fog in vicinity.

veering referring to a clockwise change in the wind direction, with respect to either height or time. Contrast with *backing*.

vertical stack The tendency for a weather system, usually a closed low or high, to have the same location aloft as at the surface. This typically indicates a lack of baroclinicity and suggests a warm-core or cold-core structure.

Vertical Totals index (VT) An expression of the low to mid-level lapse rate, as given by T_{850}-T_{500}. A value of 26 or more is considered significant.

VIL Vertically Integrated Liquid

VIS visibility.

VLCTY velocity.

VLNT violent.

VLY valley.

volume scan The complete scan of a weather radar for all assigned elevations. When a volume scan is complete, the radar is able to generate all possible products (with the exception of products that require a history of an echo). The WSR-88D completes a volume scan in 5 to 10 minutes.

vort max The highest vorticity in a given region.

vorticity The rotation in a volume of air, made up of shear and curvature.

VR veer.

VRB variable.

VRISL Vancouver Island in British Columbia.

VRT MOTN vertical motion.

VSB Visible

VSBY visibility.

VSBYDR visibility decreasing rapidly.

VSBYIR visibility increasing rapidly.

VV vertical visibility.

VWP VAD Wind Profile. A plot of winds with height above a given station, as determined by a weather radar. The profile is displayed with height as the Y-coordinate and time as the X-coordinate.

VT Vertical Totals index (q.v.)

W 1 west. 2 warm air mass.

WA AIRMET.

WAA Warm air advection

WAFS World Area Forecast System

WBZ Wet Bulb Zero (q.v.)

WDC-1 Weather Data Centers in western Europe.

WDC-2 Weather Data Centers outside of western Europe.

WDLY widely.

WDSPRD widespread.

WEA weather.

Wet Bulb Zero (WBZ) height at which the wet bulb temperature drops below freezing, expressed as height above ground level (AGL). It is a measure of depth through which a hailstone will melt. Values of less than 10,000 ft are associated with large hail, given enough instability.

WFO Weather Forecast Office

WFP warm front passage.

WINT winter.

WK weak.

WLY westerly.

WMSC Weather Message Switching Center

WMO World Meteorological Organization

WND wind.

WNWRN west-northwestern.

WNWWD west-northwestward.

WPLTO western plateau.

WR wet runway.

WRM warm.

WRMFNT warm front.

WRNG warning.

WS 1 wind shear. 2 significant weather bulletin (SIGMET) for pilots.

WSFO Weather Service Forecast Office

WSHFT wind shift.

WSO Weather Service Office

WSOM Weather Service Operations Manual.

WSR wet snow on runway.

WST Convective SIGMET for pilots

WSTCH Wasatch Range.

WSWRN west-southwestern.

WSWWD west-southwestward.

WTR water.

WTSPT waterspout.

WV 1 wave. 2 wind at altitude forecast.

WW weather watch (thunderstorm or tornado).

WWB World Weather Building

WX 1 weather. 2 flight visibility and flight weather. [PIREP]

WXCON 1 weather reconnaissance flight report. 2 weather conference.

X

XCP except.

XPC expect.

XT extend.

XTRM extreme.

XTSV extensive.

Y

Y Yukon Standard Time.

YKN Yukon.

YLSTN Yellowstone.

Z

Z Zulu Time (Greenwich Mean Time)

ZFP Zone Forecast Product (q.v.)

ZI 1 zonal index. 2 zone of interior.

Zone Forecast Product (ZFP) A NWS bulletin that provides a clear, chronological statement of the weather conditions in a county or a given set of counties for the general public.

ZRNO freezing rain information not reported.

References

References

Some cardinal references have been annotated using a. *Bracketed citations refer a technique compilation in which the original source was uncited and could not be determined at publication time.* Rules which use bracketed citations are from an unknown origin and, though published in this book as being useful, should be used with caution. The author welcomes clarifications on sources for future editions.

Official directives

It was thought appropriate to annotate official directives, all of which are recognized throughout the meteorological community, using a special tag.

FMH: *Surface Weather Observations and Reports*, Federal Meteorological Handbook FMH-1, Office of the Federal Coordinator for Meteorology, Washington D.C. (2005)

MET: *Observer's Handbook*, Fourth Edition. United Kingdom Meteorological Office, Her Majesty's Stationery Office, London. (1982)

Technique collections

Reference tags in brackets indicate a technique compilation for which the original source is uncited and could not be determined by the author at publication time. Though published in this book as being useful, these methods should be used with caution. The author welcomes clarifications on sources for future updates.

[ATC/ER]: Air Training Command, 1987: *Synoptic Systems Forecasting: Empirical Rules*. C3AAR25170 004-HO-208a.

[ATC/HDA]: Air Training Command, 1990: *Real Data Analysis*. C3AAR25170 005-HO-301B.

[ATC/SAT]: Air Training Command, 1991: *Satellite*. C3AAR25170 005-SG-117.

[EUR]: Bachman, Robert G., 1979: *A Catalogue of Some Forecaster Hints Applicable to European Forecasting*. 2nd Weather Wing Technical Note 79-004, United States Air Force.

[FRB]: Meteorological Office, 1993: *Forecasters' reference book*. United Kingdom Meteorological Office, Bracknell.

[FTW]: National Weather Service Fort Worth, c. 1978: untitled collection of empirical rules. Unpublished.

[MT]: Air Force Weather Agency, 1998: *Meteorological Techniques*, AFWA TN-98/002, 15 July 1998.

[WFR]: Elliot, George, 1988: *Weather Forecasting Rules*, American Press, 153 pp.

Original reference works

Original references, which double as suggested reading materials, are as follows:

American Meteorological Society, 1993: Guidelines For Using Color to Depict Meteorological Information: IIPS Subcommittee for Color Guidelines. *Bulletin of the American Meteorological Society*, **74**, 1709-1713.

Barthram, J.A., 1964a: A diagram to assess the time of fog clearance. *Meteorol. Mag.*, **93**, 51-56.

Barthram, J.A., 1964b: A method of forecasting a radiation night cooling curve. *Meteorol. Mag.*, **93**, 246-251.

Bunkers, Matthew J., Klimowski, Brian A., Zeitler, Jon W., Thompson, Richard L., and Weisman, Morris L., 2000: Predicting Supercell Motion Using a New Hodograph Technique. *Weather & Forecasting*, **15**, 61-79.

Burroughs, L. D., and Brand, S., 1972: *Speed of Tropical Storms and Typhoons After Recurvature in the Western North Pacific Ocean*. ENVPREDRSCHFAC Tech Paper No. 7-72, Naval Postgraduate School, Monterey.

Callen, N.S. and Prescott, P., 1982: Forecasting daily maximum surface temperature from 1000-850 millibar thickness lines and cloud cover. *Meteorol. Mag.*, **111**, 51-58.

Canadian Meteorological Center, 1987: *Meteorological Charts*, TIE Document 0055.

Corfidi, Stephen F., 2004: The Development and Movement of Haze over the Central and Eastern United States. *National Weather Digest*, Mar. 1996, updated 2004.

Craddock, J.M. and Pritchard, D.L., 1951: Forecasting the formation of radiation fog - a preliminary approach. *Met. Res. Pap. No. 624* (Met Office).

Doswell, Charles A. III, 1982: *The Operational Meteorology of Convective Weather*. NOAA Technical Memorandum NWS NSSFC-5.

Doswell, Charles A. III, 1986: The Human Element in Weather Forecasting. *National Weather Digest*, **11**, 6-17.

Doswell, Charles A. III, 1987: The Distinction Between Large-Scale and Mesoscale Contribution to Severe Convection: A Case Study Example. *Weather and Forecasting*, **2**, 3-16.

Doswell, Charles A. III, 1992: Severe Local Storms Forecasting. *Weather and Forecasting*, **7**, 588-612.

Doswell, Charles A., Johns, Robert H., and Weiss, Steven J., 1993: Tornado Forecasting: A Review. *The Tornado: Its Structure, Dynamics, Prediction, and Hazards*, C. Church et al., Eds., Geophysical Monograph 79, Amer. Geophys. Union, 557-571.

Doswell, Charles A., 1999: *Vorticity Advection and Vertical Motion*. <http://www.cimms.ou.edu/~doswell/PVAdisc/PVA.html>

Dvorak, Vernon F., 1975: Tropical Cyclone Intensity Analysis and Forecasting from Satellite Imagery, *Monthly Weather Review*, Vol. 103, pp. 420--430, 1975.

Dvorak, Vernon F., 1984: *Tropical cyclone intensity analysis using satellite data*. NOAA Tech. Rep. NESDIS 11, National Oceanic and Atmospheric Administration, U.S. Department of Commerce, Washington, DC, 47 pp.

Dunn, Lawrence, 1987: Cold Air Damming by the Front Range of the Colorado Rockies and its Relationship to Locally Heavy Snows. *Weather and Forecasting*, **2**, 177-189.

Dunn, Lawrence, 1991: Evaluation of Vertical Motion: Past, Present, and Future. *Weather and Forecasting*, **6**, 65-75.

Federal Aviation Administration, 1999: *Aviation Weather Services*, AC 00-45E, U. S. Department of Transportation.

Felsch, Peter and Whitlatch, Woodrow, 1993: Stratus Surge Prediction along the Central California Coast. *Weather and Forecasting*, **8**, 204-213.

Ferman, M. A., Wolff, G. T., and Kelly, N. A., 1981: The Nature and Sources of Haze in the Shenandoah/Blue Ridge Mountains Area. *Journal of the Air Pollution Control Association*, **31**, 1074-1082.

George, John E. and Gray, William M., 1977: Tropical Cyclone Recurvature and Nonrecurvature as Related to Surrounding Wind-Height Fields, *Journal of Applied Meteorology*, **16**, 34-42.

Gerhardt, J. R., 1962: An example of a nocturnal low-level jet stream. *J. Atmos. Sci.*, **19**, 116-118.

Glickman, Todd S., Macdonald, Norman J., and Sanders, Frederick, 1977: New Findings on the Apparent Relationship Between Convective Activity and the Shape of 500 mb Troughs. *Monthly Weather Review*, **105**, 1060-1061.

Golding, B., 2002: *Met Office forecasting methods for fog and low cloud.* Met Office slide presentation. < http://lcrs.geographie.uni-marburg.de/fileadmin/COST_media/WG1/helsinki/UK_Golding__MetOffice.ppt >

Henry, W. K., 1949: *On the Movement of the Southwest Low.* Thesis for the degree of Master of Science, University of Chicago, September, 1949 (unpublished).

Holton, J. R., 1972: *An Introduction to Dynamic Meteorology.* 1st ed., Academic Press, 319 pp.

ICAO, 2006: *Location Indicators*, Doc. 7910/119, March 2006. International Civil Aviation Organization, 236 pp.

Jefferson, G.J., 1950a: Method for forecasting the time of decrease of radiation fog or low stratus. *Metorol. Mag.*, **79**, 102-109.

Jefferson, G.J., 1950b: Temperature rise on clear mornings. *Meteorol. Mag.*, **79**, 33-41.

Johnson, D.W., 1958: The estimation of maximum day temperatures from the tephigram. *Meteorol. Mag.*, 87, 265-266.

Little, C. D., 1985: Isentropic Plotter. *NOAA Eastern Region Computer Programs and Problems*, NWSA ERCP No. 29, NOAA, U.S. Department of Commerce, 10 pp.

Macdonald, Norman J., 1976: On The Apparent Relationship Between Convective Activity and the Shape of 500 mb Troughs. *Monthly Weather Review*, **104**, 1618-1622.

Maddox, R. A., 1976: An evaluation of tornado proximity wind and stability data. *Monthly Weather Review*, **104**, 133–142.

Masterton, J. M. and F. A. Richardson, 1979: *Humidex, a method of quantifying human discomfort due to excessive heat and humidity.* Environment Canada, 45 pp.

McKenzie, F., 1943: A method of estimating night minimum temperatures. *Met. Office, Synoptic Div. Tech. Memo. No. 68* (unpublished document).

Moore, James T. and Smith, Kenneth F., 1989: Diagnosis of Anafronts and Katafronts. *Weather and Forecasting*, **8**, 61-72.

Namias, J., 1940: *An Introduction to the Study of Air Mass and Isentropic Analysis*. American Meteorological Society.

Oard, Michael J., 1993: A Method for Predicting Chinook Winds East of the Montana Rockies. *Weather and Forecasting*, **8**, 166-180.

O'Lenic, Edward, 2004: The Long Journey of NWS Medium Range Prediction. *Symposium on the 50th Anniversary of Operational Numerical Weather Prediction*, College Park, Maryland, June 14-17, 2004.

Price, W. B., 1971: *Wind and Weather Regimes at Great Falls, Montana*. Western Region Tech Memo No. 64., National Weather Service. 64 pp.

Rekacewicz, Philippe, 1998: *AMAP Assessment Report: Arctic Pollution Issues*. Arctic Monitoring and Assessment Programme, Oslo, Norway.

Riehl, H., and Shafer, R. J., 1944: The Recurvature of Tropical Storms. *Journal of Meteorology*, **1**, 42-54.

Sansom, H. W., 1951: A study of cold fronts over the British Isles. *Quarterly Journal of the Royal Meteorological Society,* **77**, 96-120.

Saunders, W.E., 1950: A method of forecasting the temperature of fog formation. *Meteorology Magazine*, **79**, 213-219.

Shaw, Glenn E., 1995: The Arctic Haze Phenomenon. *Bulletin of the American Meteorological Society*, **76**, 2403-2413.

Siple, P. A., and C. F. Passel, 1945: Measurements of dry atmospheric cooling in sub-freezing temperatures. *Reports on scientific results of the United States Antarctic Service Expedition, 1939–1941*. Proceeds of the American Philosophical Society, **89**, 177–199.

Snellman, Leonard, 1986: *Use of Isentropic Charts in the 1980s*. Western Region Technical Attachment 86-09, National Weather Service.

Steadman, R.G., 1979: The assessment of sultriness. Part I: A temperature-humidity index based on human physiology and clothing science. *Journal of Applied Meteorology*, **18**, 861-873.

Strauss, Harald, 1979: *A New Technique for Forecasting the Occurrence of Fog and Low Stratus Ceilings by Use of Chart.* 2WW TN/79-008, U.S. Air Force, 35 pp.

Trenberth, K. E., 1978: On the Interpretation of the Diagnostic Geostrophic Omega Equation. *Monthly Weather Review*, **107**, 682-703.

Velden, C. S., 1990: The impact of satellite-derived winds on hurricane analysis and track forecasting. *Preprints, Fifth Conf. on Satellite Meteorology and Oceanography*, London, England, Amer. Meteor. Soc., 215–219.

Wantuch, Ferenc, 2002: Visibility and Fog Forecasting Based on Decision Tree Method . *COST Action 722: Short Range Forecasting methods of fog, visibility, and low clouds.*

Weber, Eugene, 1980: *Winter Patterns*, 3WW/FM-80/011 Forecaster Memo, U.S. Air Force, 26 pp.

Weber, Eugene, 1981: *Satellite Interpretation*, 3WW/TN-81/001 Technical Note, U.S. Air Force, 95 pp.

Wilson, L. J., 1985: *Isentropic Analysis - Operational Applications and Interpretation*, third edition. Edited for Training Branch by James Percy. Atmospheric Environment Service, Canada.

WMO, 1975: *Vol I: Manual on the Observation of Clouds and Other Meteors*, revised edition. World Meteorological Organization Pub. No. 407, Geneva.

Wolff, Paul M., 1955: Quantitative Determination of Long Waves and their Time Variations. *Journal of Meteorology*, December, 1955, **12**, 536-541.

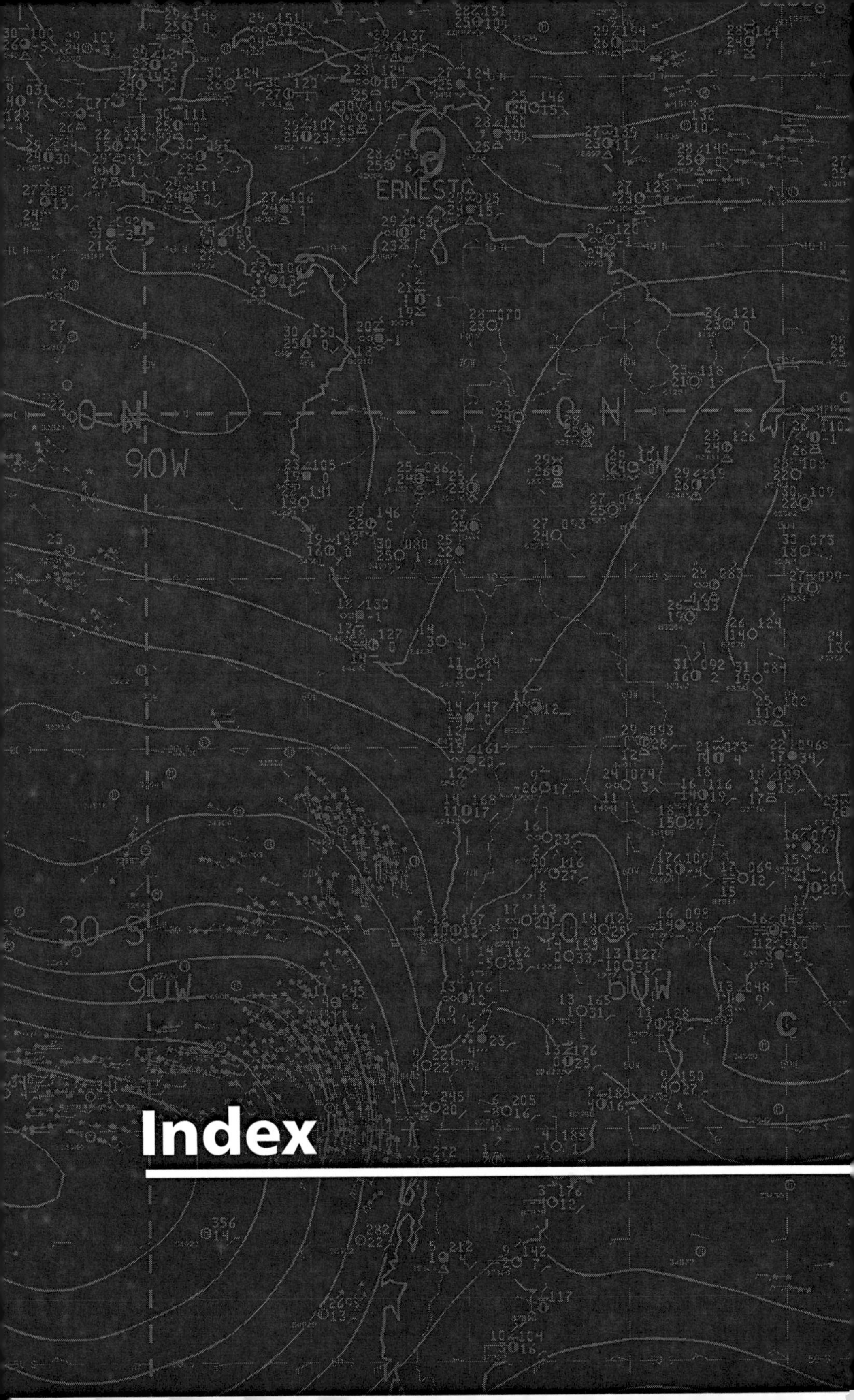

Index

Index

Symbols

A

B

C

D

E

CPSIA information can be obtained at www.ICGtesting.com
Printed in the USA
LVOW05s0537060214

372596LV00002B/119/P

9 780970 684066